THIS IS A SEVENOAKS BOOK

This edition published in 2004 by
Sevenoaks
An imprint of the
Carlton Publishing Group
20 Mortimer Street
London
W1T 3JW

A CIP catalogue for this book is
available from the British Library.

ISBN 1 86200 173 1

Project Manager: Stella Caldwell
Editor: Peter McSean
Design: SMITH
Picture researcher: Steve Behan
Production: Marianna Wolf

SEVENOAKS

THE CAR

A HISTORY OF THE AUTOMOBILE

JONATHAN GLANCEY

JONATHAN GLANCEY IS ARCHITECTURE AND DESIGN EDITOR OF *THE GUARDIAN*. A FREQUENT RADIO AND TV BROADCASTER, HE IS AUTHOR OF *THE STORY OF ARCHITECTURE* AND CARLTON'S BEST-SELLING *C20TH ARCHITECTURE*. A JAGUAR FAN SINCE THE DAY HE FIRST SAW ONE OF THE LAST NEW MK2s HOWLING OUT OF A LONDON SHOWROOM, HE HAS ALSO OWNED AND COPED WITH A VARIETY OF BRITISH CLASSIC CARS. HE LONGS FOR AN ASTON-MARTIN DB4GT.

GET IT AT Whelan
DRUGS COSMETICS
SODA PRESCRIPTIONS
MASTER REALTY
REAL ESTATE-INSURANCE
CHASE MANHATTAN BANK
CHASE MANHATTAN BANK
16 Lessons $10
Whelan
CUT RATE
Drugs
SODA
UNITED CIGARS
TELEPHONE BOOTHS
BAR & GRILL
BAR & GRILL
STEAKS CHOPS
PUBLIC PARKING AREA
IRT

Beneficial
Loans
NITED
SUBWAY
IRT
CHASE MANHATTAN BANK
Q-13
FLUSHING
BAYSIDE

PLK 569

INTRODUCTORY ESSAY

A
EGYPTIAN WAR CHARIOT
Faster than fairies, faster than witches; if not quite as fast as the sky gods, this was the way to travel 3,500 years ago: ecologically sound, lightweight, reliable, fast, and with plenty of fresh air...what more could any living god want?

B
LOVE ME, LOVE MY CAR
Are these chaps waking her up or putting her to bed? Either way, we tend to cherish the car more than we do our families, loved ones and animals.

C
TAKE THIS!
Henry Ford was in no mood for pleasantries when he set about one of his cars – not because he detested it, but because, in this case, he wanted to prove how tough it was.

A

Tutankhamun died aged 19. The teenage pharaoh was buried in an underground tomb in the Valley of Kings at Luxor in southern Egypt. When this was discovered some 3,000 years later, by Howard Carter in 1922, a craze for all things ancient Egyptian was nurtured among European and American fashion designers, set designers, jewellers, Hollywood scriptwriters, authors and architects. The discovery of this fabulous desert treasure trove coincided, more or less, with the great exhibition of decorative arts held in Paris in 1925. This gave birth to Art Deco design, which incorporated much ancient Egyptian design drawn from the finds in Tutankhamun's tomb. Within a couple of years, buildings and ball gowns, ocean liners, express trains and, yes, even cars went Deco crazy, the Chrysler building as well as Chrysler cars. The spirit of Tutankhamun was with them.

And yet, for all their fancy dress, many Art Deco buildings and very many Art-Deco-adorned automobiles were less sophisticated in a number of engineering ways than six very special machines found by Carter in the Valley of the Kings a little over 90 years ago. These six machines were Tutankhamun's war chariots. The first time I saw them, some time in the late 1980s, in Cairo's magnificent Egyptian Museum, I was as dazzled by their construction as by their decoration. Not only did the chariots appear to be brand new, but they were distinctly modern in a way that Colin Chapman, founder and designer of Lotus Cars and its lightweight, nimble, sporting machines, would have appreciated thousands of years on. True, my first impression was that these evidently rapid machines were the Ferraris of their day; or Ferraris as if styled by Versace. Yet the more I gazed at them, the more I came to look beneath the surface gilt. And, now, I know – courtesy of a number of American academics – that Tutankhamun's chariots really were high-performance machines. Aside from ultra-lightweight superstructures, they boasted lubricated, long-life bearings, tunable, semi-independent suspension, multi-layer tyres, a structure designed to withstand bending and twisting, and spoked wheels that would have given these machines and their riders an exceptionally stable ride. With just 2hp up front, the chariots would have been faster than any competitors. Written records testify to their success in battle.

So here were high-performance machines of their day that, in certain ways, were more sophisticated than an Art Deco Chrysler, for all its beefy engine, pneumatic rubber tyres and plush, sofa-style seats. I remember thinking on that first encounter with Tutankhamun's chariots that I would have preferred to get around by these machines at the turn of the twentieth century than by one of Henry Ford's first cars. They would have been faster, more reliable, more efficient in energy use, kinder to the environment and infinitely more stylish.

That day, I had come to the museum, in one of those awkwardly stretched Mercs you find across the Middle East, from the pyramids at Giza, buildings infinitely more refined and sophisticated than many of the Post-Modern blunders blighting the world's cities in the 1980s. To say the road was busy was like saying Harrods sale is a bit of

B

a crush. Overladen lorries, smoking coaches, pick-ups of every kind sashayed in a drunken dance along the die-straight avenue that leads from Giza to Cairo. Between them, donkeys borne down with panniers, rib-thin horses pulling carts at a trot, occasional camels, mad dogs – and this Englishman. It was very hot. A donkey pulling an abusively heavy cart collapsed in front of my cab. The driver, in full view of his family and the crawling traffic, jumped down from his seat and began lashing the pathetic beast mercilessly with a cruel whip.

I jumped out and, after a brief discussion, clocked the driver, and rested the animal's head in my lap. Water was brought by a moved, or guilt-ridden, coach driver. The police arrived. I was accused of assault. I was, it has to be said, ashamed of my aggression – I am normally a mild-mannered fellow – but my side, thankfully, was taken by the crowd. The donkey, foaming at the mouth and shaking violently, gave a huge sigh, and expired. I unhooked reins that been digging into what flesh covered its old bones. The cart it pulled had all the sophistication of a 1970s Morris Marina: slow, resisting, uncomfortable, wobbly, hard to steer. The animal had been forced to work much harder per kilo carried than the lithe war horses that would have raced Tutankhamun along antique roads and across ancient dunes.

Several thoughts crossed my mind. Would the pugnacious cart driver have beaten a pick-up if it had broken down on the Giza Road? Would he have run it on tyres inflated with about as much air as a canary carries in its tiny, trilly lungs? Might he have paid attention to the gauges on the dashboard of his vehicle and noticed that something was awry? He clearly didn't notice – but, then, I suppose, he didn't care – that his donkey was labouring fatefully and in danger of imminent collapse.

Yes, John Cleese, in the guise of Basil Fawlty, the enraged hotelier star of the classic BBC TV comedy "Fawlty Towers" did thrash his BMC 1100 when it failed to start at a time when he most needed it to save his professional skin, but relatively few motorists are as idiotic as Basil Fawlty or as stupidly cruel as my Egyptian cart driver. Like you, I have witnessed people who wax, polish and T-Cut their cars even as they kick the dog, curse the cat, and generally abuse their families.

Why is this? Because the mechanically propelled vehicle, and most of all the car, is a tin god, and one of our own making. Since its invention proper in the 1890s, in Germany and France, the car has been worshipped like no other machine. More temperamental than a Hollywood star, greedier than a Wall Street financier, dirtier, when all is said and done, than a pig rolling in mud, and more destructive than a fireworks factory in a thunderstorm, the car is a curious bedfellow. Today we find it almost impossible to live without it. We fuss over it. Talk about it proudly in pubs. We build it little houses to live in. It is truly part and parcel of our lives, every bit as necessary to us as dogs were to Stone Age hunter-gatherers and settlers alike. Or horses to young Egyptian kings.

Car factories themselves are becoming increasingly like places of worship, or art galleries, which are much the same thing. Just look at Volkswagen's latest glass

C

D
AUTOMOTIVE NYMPH
Young woman – a virgin, no doubt – sacrificed on the altar of the great god, Automobile. We worship this machine to an unhealthy degree.

E
SPOT THE ROVER
The car quickly became an intimate part of our lives. Births, funerals, weddings: it plays a part in all of them.

D

assembly plant in Dresden, or the new Rolls-Royce factory on the Goodwood Estate, home of the annual Festival of Speed, in West Sussex. Cars themselves are becoming ever more lavish as our love for them breeches the wall of common sense. Our cars are becoming mobile second homes and alternative offices. We are willing to sit in Giza-Road-style traffic jams because our latest loves boast air-con, quadrophonic sound, mobile phones, computer games, radio and even TV. "Please, Herr Glancey, do not attempt to watch the television while you are driving," implored a press lady from BMW's Munich headquarters some while ago. I had been joking. Lent a glorious, blood-red, six-speed, 400bhp M5 saloon to complete a tour of Bavaria's baroque and rococo monasteries, I had called BMW on my way back to Munich, confused by the profusion of autobahn exits. The car rode magnificently at any speed. The press lady asked me if I was experiencing any problems with the car. No, I said, except that I found it hard to watch the comedy on TV while overtaking at 240km/h… "Please, Herr Glancey, do not attempt to watch the television while you are driving."

That particular car had transported me effortlessly from house of God to house of God. It was in itself an almost sacred idol, worshipped by Bavarians – and many others around the world – who know and appreciate a great car when they see one. They made this one, too, along with the miraculous churches of their picture book Alpine state.

In motor shows worldwide, worshippers come in tens of thousands to pay homage to new gods, as if propitiating the latest offspring of Zeus. Otherwise healthy young women, automotive nymphs, are strewn all but naked, like sacrificial victims, across the lovingly polished machines. They pout, writhe and open and close their mouths as in some sacred sexual frenzy. The faithful can only drool as they worship from afar. They are not allowed to approach too close to the altar of automobilia, much less to touch either the nymphs or the fleet-footed messengers of the gods named, variously, Aston Martin, Ferrari and Lamborghini.

On a much humbler level, I remember from childhood when we used to collect for a charity called Princess Elizabeth Day. The money went to poor and, presumably, car-less children. In one suburban house, gleaming from damp-proof course to chimney top, lived an immaculate lady we called Mrs Godhouse, because she cleaned her house continually and had plastic runners placed the length of her hall to stop the carpet from getting dirty. One year, we came with our collecting box to find that Mr Godhouse had taken delivery of a new car, an Austin 1100 like Basil Fawlty's. From the day the Issigonis-designed, front-wheel-drive runabout arrived, the plastic seat covers that protected the new car's shiny vinyl seats stayed put. The car, like the house it sat outside, was holy. To sit on its seats without protection would have been anathema.

Cars in general, though, are anathema to many conservation-minded people today. When I went to interview Sir Wilfred Thesiger, the legendary explorer and travel writer, last year, he told me that he feared the car. Yes, he had once owned a Land Rover in his years

E

living in Kenya, but he had walked most of his epic journeys and now saw the car as one of the greatest dangers threatening civilisation and humankind. I happen to think Sir Wilfred is more than probably right. Like God, the car gives even as it takes away. It offers a spurious freedom while jamming up our cities and siphoning off our natural energy reserves. It promises a dream that it cannot deliver. Only rarely can most of us enjoy the freedom of the road known to Tutankhamun and dreamed of by Henry Ford, along with Mr Toad and legions of romantic race and rally drivers. Given the density of modern traffic and the encylopaedia of regulations governing road use, we could never race an express train from the Cote d'Azur to Paris as a vintage Bentley did the Blue Train in the 1920s.

Within a few years, millions of proud new car owners in the developing world, among them my Egyptian donkey driver I have no doubt, will take to ever more roads. Landscapes will be dug up to create a subtopia designed so that we journey from banal shopping malls to cynically built housing estates, via theme parks, schools, air-conditioned office blocks and banal business parks, buildings and experiences created and served by the car. In southern China, I have witnessed the demolition of entire hills, and their flora and fauna, to make way for new roads, factories and housing estates designed on lap-top computers by zealous, young, business-minded technocrats.

To say "I don't own a car" is tantamount to being a loser in the modern, ever-mobile world. To go further and say "I don't want a car" is conceived of as eccentricity or even perversity. In Britain, if you do not own or want a TV, the licensing authorities will keep bothering you, threatening you with penalties and punishments if you do not buy a TV licence. Why? Because those who run the licensing authority cannot believe that there are people in twenty-first century Britain who would prefer to read, walk the dog, cook, make something, or just dream the day away, rather than sit, stupefied, in front of a demanding little box gawping at tripe between adverts for even more tripe. And for cars.

Here is the great abyss that car lovers with any sensitivity to life, love, sanity, their own and other peoples' cultures, for other species, including Giza Road donkeys, and to the planet itself try not to stare into too hard or often when they slip behind the wheel of their latest automotive dream. The car is both demi-god and devil, a temptation beyond endurance, a blight, a curse, a disease-riddled whore... but also a prized, if jealous, lover. The car purrs into our ears, promising a dream of open roads, wind-in-the-hair motoring, appealing to the latent Stirling Moss, Michael Schumacher or, even, Tutankhamun in so very many of us.

The car bug – autophilia – is at least as old as ancient Egypt. Just as humans yearned to fly, so they longed for a form of transport that would propel them at speed across desert, steppes and savannahs. Of course, they also wanted sturdy and reliable vehicles to shoulder goods across these same landscapes so that they could earn the kind of money – surfeit, profit – that would, one day, enable them to indulge in luxuries that, in

F
HORSELESS CARRIAGE
It took a while for the car to develop an aesthetic all of its own. It is easy to imagine horses pulling a handsome, craft-built carriage like this.

G
THE PROMISE OF FREEDOM
The car promises more than it can ever really deliver; where can you drive a car like this E-Type, as it was designed to be driven? In your head, in your dreams.

F

Tutankhamun's day, included the nimble, lightweight, war chariot.

Oxen, llama, camels, yak, donkeys and other beasts of burden performed, and still perform, the work of today's lorries and vans. Only horses, though, could really get us going fast. But a fast horse is a pedigree creature; it has never, except in military and certain nomadic societies, been a form of transport for the common man, much less woman. Speed, then – which along with the notion of independence, is the great attraction of the car, as opposed to a lorry or van – was a mercurial realm in which only the rich, their favourites, or the gods might have the opportunity to play.

Even when the first cars emerged, coughing, spluttering and frightening the horses little more than a century ago, they were sluggish things. Almost any horse could outpace every one of the earliest horseless carriages. In any case, these comical-looking gadgets had none of the beauty or elegance of line, much less the dignity of a pedigree horse. And it was to be very many years, as this book shows, before the car developed an aesthetic of its own, separate from that of the horse-drawn or railway carriage.

The first challenge had been to find a reliable and lightweight means of motive power that would allow an automobile to out-distance a horse. Several well-recorded experiments were made over the centuries, even the millennia, with forms of steam power.

Five hundred years before Christ rode a donkey into Jerusalem at the height of his human fame, Hero of Alexandria demonstrated a steam turbine. It was much admired, but no one really knew what to do with it. After all, what was the point of a steam-powered warship when a man and sail-powered trireme or quinquireme could go as fast as anyone needed to go at the time? Facts which had made sea-going, horse-worshipping ancient Athens a regional superpower? Which king would have needed so much as a steam pressure-cooker in an age when he could engage legions of slaves to cater for his every whim?

Moving quickly on, which eighteenth century king, queen or emperor would have wanted to travel by Cugnot's ungainly steam carriage? This clumsy brute hissed, smoked, generated dirt and, in any case, had a top speed, in a short sprint, of 4mph. Even a silken princess at the court of Versailles could move faster on her own two dainty feet.

The car proper finally emerged with the development of the internal combustion engine. This was at a time when expresses on Britain's railways were well able to run safely and smoothly at 90mph and could cover long distances, with handsome dining cars and lavatories, at the rate of a-mile-a-minute. It took those first faltering cars some while to catch up with steam expresses, but when they did, they promised to change the face not just of public transport, but of the world.

Once its reliability was proven, the car spelt freedom. Of course, it took the development of motor roads, enduring tyres, petrol stations and workshops to help it on its way, but soon enough it was spinning all day, every day, the length and breadth of continents, a machine for all seasons and purposes. As this book shows, the car blossomed to become an integral part of our lives. It changed the face of roads and the housing and architecture that grew up alongside these. It created the motorway, the garage, the petrol station, motels, shopping malls, drive-in cinemas, much of suburbia, and traffic jams. It helped police and criminals alike. It gave us drive-by shootings and the armoured car. It gave us reels of Hollywood cop movies and car chase sequences. It became a prop for film and TV stars. It gave horses a race for their money as grand prix circuits flourished. Above all, it became both an inevitable form of transport and a highly desired status symbol.

We all know, though, that the freedom Henry Ford promised us is unreal. It might be real, kind of, if you happen to live in rural Montana, the heart of Canada, in the middle of Australia or in a Germany still free of a maximum speed limit and where cars are engineered so thoroughly, expensively and well because they really do get driven, day-in, day-out, at sensational speeds. For the rest of us, though, the quality of our car, the state of our roads, laws or sheer congestion conspire to slow us down, hem us in. There may be nowhere to park once you have left home. It might well be better to commute by crowded commuter train than by car, yet millions of drivers still set out each morning from their homes in the knowingly mistaken belief that today's journey will be better than yesterday's, that, somehow, the tide of traffic will part as miraculously as the Red Sea did for Moses, and that we will drive heroically, speedily and unimpeded just as smooth-looking models do in cinema and TV ads.

Oddly, this really did happen to me one day. I was driving a brand new V12 Sovereign from Jaguar's Browns Lane factory back home to London. Whichever way I nosed this silent saloon, the main roads were choc-a-bloc. I turned off the first minor road I could and threaded towards Banbury to look at the fine Neo-Classical church at the top of the hill there, and for a cup of tea and Banbury cake, before riding, fresh like a cock-horse, on my rural way. I made a turn when no other car turned and the road seemed to disappear. It was a Friday early evening in June. I threaded the big, sure-footed car through strange chicanes. There was no traffic. And suddenly before me was a great open road without a single car, much less a donkey and cart or king's chariot, anywhere in sight. Cautiously, I launched the Jag down on to this miraculous freeway, and opened her up. Fifty, sixty, seventy miles per hour. Nothing. Just an unimpeded sweep of uncannily smooth concrete. I crested the brow of a hill. Still no other road user in sight. Summer swifts darting and a hovering kestrel above, but absolutely nothing on the road. I was off. Foot sunk gently but deeply into the Wilton pile and the big-hearted car gathered speed like Concorde on its way (no more, sadly) from Heathrow to JFK. Down the hill I peaked out, "maxed" the car as they say in machismo real men's car mags (along with "stump-pulling torque" and other gym-pumped terms). The speedometer read 150mph, the rev-counter needle haunted the red mark on its dial. Sheer, Toad-like bliss. I eased up comfortably in time to see barriers across the end of my secret, dream-like road. I steered off through more curious

G

H
ASTON MARTIN ULSTER
As much fun as motoring gets: a small and potent car that feels alive without being absurdly powerful.

I
MORRIS MARINA
Fifty miles per hour is quite enough sometimes.

J
MINI 850
Some cars are designed to be fun to drive with just 34bhp under their tiny bonnets. The original Mini was spirited and fun, even if it was unprofitable and tended to fall to bits.

K
RILEY 1.5
The virtues of simple motoring. I owned one of these Rileys as a runaround for some years. It never failed to start at the first turn of the key. It never broke down.

I

chicanes and found myself lolloping up to old A-road that runs from Banbury to Oxford, birthplace of Morris and MG cars. I had been driving south along the all but completed M40 that now trawls its laboured way, stuffed with traffic, from London to Birmingham, by-passing Oxford and Banbury with its famous cakes, cross and church.

That brief spell in middle England at two-and-half-miles-a-minute is, perhaps inevitably, the fastest I have ever driven in Britain (200mph in Italy, cheered on by the *carabinieri*, in a Ferrari F40 between Modena and Florence, but that is another story told in *The Sunday Times* colour supplement much the same year I went to see Tutankhamun's fleet chariots in Cairo). It was a piece of luck, an adventure, and presumably could never happen again. Today, like many car enthusiasts in Britain, I feel ambivalent about both driving and the car. It is, all too often, no fun to motor on our crowded roads. And yet, when the chance comes to drive a nippy sports cars, no air-con, windows or roof down, across the wilds of central Wales, the Border Country between England and Scotland or some of Scotland's finest routes, the sense of pleasure and adventure returns. Even if I know I am driving in a fool's paradise.

Speed itself is relative. Politicians, apart from cabinet ministers on official business, nanny us into believing it is wrong at all times. It is hard not to agree with them when a callous young moron, all pimples, sneer and baseball cap, hurtles a deafening car heedlessly along residential and town centre streets, or careers sideways on a country road, unable to cope with the speed he has induced hormonally rather than intelligently. Because such gormless nitwits will always be with us, speed bumps, speed cameras, speed gaps and every other form of speed restriction are inevitable and will gradually whittle away at our freedom to drive as we would like to.

There are remedies. You could always choose a slow(ish), but well-engineered classic car and enjoy driving more slowly than you have become used to. An Aston Martin Ulster at 60mph offers much the same thrills as an Aston Martin Virage, 70 years its junior, going twice the speed, and with no fear of losing your licence in a moment of full throttle madness. Or you might opt for a racing cycle, or motorbike, or give up altogether and get the train. But not perhaps in Britain, where the privatized railways are more concerned with golden handshakes for directors with no interest in trains, than for their unfortunate passengers.

Astons aside, some of the most commonplace cars of the past remain fun to drive. On backroads at least. I have always enjoyed, when I have had the chance or owned them, driving, among others, Morris Minors, Citroën 2CVs, Alfasuds and Minis. The Minor offers positive steering, a clickety-click gearchange, surprisingly good handling and a jolly, burbling exhaust note that somehow suggests either it or you are in a perpetual search of slap-up cream teas with lashings of ginger beer in a village complete with duck pond, forge and cycling vicar (in search of tea, rock-cakes and antimacassars). The Deux-Chevaux – as many horses as Tutankhamun would have had to power one of his chariots – boasts one of the sweetest-singing of

H

J

K

all internal-combustion engines and more eccentricities than the late Peter Sellers and Spike Milligan combined: windows that flap open like Dumbo's ears in the slipstream and stay open at speed; doors that bulge out like sails at top speed (not quite 70 on the level, a mind-blowing 80 downhill); and a gearchange that could outfox Mr Rubick, inventor of that puzzling cube. It yaws around corners in comic fashion, yet firmly refuses to fall over. Its engine's capacity is just 602cc, yet it can carry five people and their luggage over fields as well as roads. It boasts a full-length sunroof. It is immense fun.

The noise in the cabin of my 1275cc Mini Cooper S was a particularly effective speed deterrent. Seventy was truly the limit. It drove, though, like a go-kart. Its siblings won the Monte Carlo rally, starred in the original version of *The Italian Job* and were dolled up in pyschedelic colours for such inspiring 1960s stars as John Lennon. The Neopolitan Alfa was superb – Alfa Romeos had traditionally been made in Milan – except for one small detail: when you put your foot down, it could well go straight through the flimsy floor. It attracted rust like a celebrity diner at The Ivy attracts paparazzi.

Which is where dependability and safety come into the picture. Of my own cars, a 1966 Mini 850 once shed its engine; the cheap and cheerless engine mounts gave way and the venerable A-series engine shot into the radiator. Before it was impeccably restored by Nick Goldthorpe in Bridgnorth, my 1963 3.8 Mk2 Jaguar was, very probably, the most moth-eaten mechanical cat on the road. It was held together, in various parts, with a Fanta can (exhaust), a fanfare of jubilee clips and black tape around the expansion bottle connected to its defiantly modest radiator. A modest 1965 Riley 1.5 proved to be the most reliable car I have ever owned. A 1967 Sunbeam Alpine had been fitted with a wooden piston. My V12 Sovereign was a delight, although the way the fuel gauge appeared to be permanently faulty – could the car really be using all that four-star? – had me, first, looking for a fuel leak (none) and then selling it on to a sympathetic publisher who now owns two big Jags. He must be doing well. Or else had shares in the Iraqi oil industry before anyone else.

Most cars are pretty reliable today. My Mk2 Jag's manual opened with a maintenance section headed "Daily". Check oil, water, tyre pressures etc. These things were not to be ignored. Now, like you, I know many people who never lift the bonnet of their car. Sure, they will have the thing serviced at more or less regular intervals, but the mechanical workings of their machine are taken for granted. In any case, if it fails to start, or runs low on oil, a trailer comes to pick it up. In developed countries, knowledge of car mechanics falls in relation to rising car sales, and, of course, incomes. Significantly, perhaps, manufacturers have begun to cover up their engines totally, so that if you do care to raise the hood, you are likely to meet with a silver or slate-grey screen set across the top of the engine block. This gives the impression that the car is somehow run by solid-state electronics rather than by a complex whirring and entanglement of pistons, valves, connecting rods, camshafts, crankshafts, belts,

L
1915 ROLLS-ROYCE SILVER GHOST
Form follows function; the motor car in direct and elemental form, free of power-assisted this, computer-controlled that.

M
1958 FORD PREFECT
The baby is well equipped with bottle and potty, in case of accidents. But no seat belts or kiddy seats in this more innocent, if more dangerous age.

cogs and chains, topped off with radiators, hoses, spinning turbochargers, whistling superchargers, heating ducts, batteries and all manner of gizmos and eccentrics. The last may or may not refer to the owner.

Only enthusiasts delight in getting their hands oily, while service mechanics are paid through the nose to do so. To the majority of owners, the way a car works is increasingly the stuff of Byzantine courts: secret, inscrutable, ineffable. Significantly, too, few manufacturers this side of the sporting or machismo enthusiasts' market sell their cars on their technical specification any longer. Advertising is concerned with being, first, clever and ironic, and, second, with "lifestyle", that improbable string of brands and labels we are meant to hang around our supposedly sophisticated necks to prove just how smart, cool, with-it, discerning and wealthy we are. But, actually, just a bit naive and all too willingly taken in by hype.

Recently, I got to drive one of those funny, upright little Audi A2s, a breed of mechanical terrier, that I took to like a duck to a gamekeeper's gun. It has a panel that you can lift to top up the radiator, should you ever have to. Owners need never look at the engine, wherever it is, ever. Engines in older generations of cars were designed not to be ignored but to be looked at and admired. The engine cowlings of Rolls-Royces, Mercedes-Benzes, Hispano-Suizas and those of other grand, double-barrelled machines were designed to open up at the side, so that the great motors, all shining brass, steel and aluminium, could be revealed in all their might and magnificence. These were engines that appeared to have had more in common with the great workings of *The Flying Scotsman* or the *Queen Mary* than with modern cars, ashamed of their venerable mechanical underpinnings. In a digital age, who cares for pistons and connecting rods?

One of the great delights of cars built before the electronic and digital ages is that everything in them is as free from power assistance and digital gizmos or, in other words, as genuine, straightforward and somehow as right as possible. I have to be careful here. I don't mean sit-up-and-beg Ford Prefects or any number of slow and creaky cars no more sophisticated than the Mamod steam engines with which I filled the house up with saturated steam, spitting oil and the strangely intoxicating fumes of methylated spirits as a child, but the likes of Mercedes 540Ks, Bugatti 35Bs, Deusenbergs and Bentley Speed Sixes. Those cars had an honesty about them, a mechanical sincerity and, by our standards, an engineering other-worldliness. Having had the chance to drive these machines, each has been a revelation. The Bugatti remains sensational, an all but peerless racing car in the 1920s, a very fast car today, as sensual to pilot along the road as a Spitfire is in the sky.

As for safety, things have changed, and very much for the better. And yet, there is something sad, if not sane, in the loss of motoring innocence. The joys of sitting on fathers' knees as they smoked untipped cigarettes, quarterlight windows open, and you steered and changed gear, and asked, politely, if you could go a hundred, are very much gone. Unquestionably seat belts, airbags, power-assisted steering, powerful brakes, children's

L

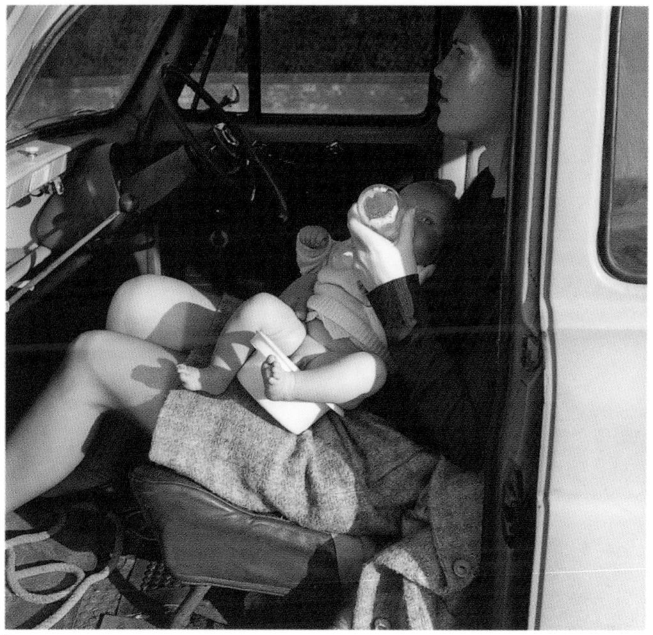

M

seats, speed limits and other devices inside and outside the car have saved many lives. Perhaps little Joshua and Jessica are a lot safer being driven to expensive nursery schools two miles down the road in a Mitsubishi Shogun, Range Rover or Jeep Grand Cherokee protected with bull-bars than walking or taking the bus or train, where at least every second filthy, disease-infested seat is taken by murderers, molesters, pyschotics, perverts and other penny-dreadful tabloid monsters.

A couple of years back, I witnessed a fascinating cameo of two families at play on the wonderfully blustery and unspoiled beach at Holkham in North Norfolk. Three bright and beady local children were playing with their boisterous mongrel. They were barefoot and barely dressed even though it was cold. They were having great fun. They happened to run back to the car park at the end of the beach's long boardwalk as I jogged back with my dog. Dad, in T-shirt and jeans, was reading a tabloid paper behind the wheel of a sand-flecked Toyota pick-up. "In yer get," he hollered without looking up. Up bounced the dog, in clambered his playmates among the various boxes in the back of the van, and off they went chattering and barking into rural Norfolk. Perhaps a passing safety inspector or policeman would have had a fit. I do not know. In any case, my attention was drawn to a sobbing boy, of much the same age as the Toyota gang, dressed in layers of posh rain-proof clothing and brand new wellingtons. He was being strapped tightly into a rear-facing child's safety seat of an air-conditioned, top-of-the-range Jeep by a no-nonsense smokey-blonde matriarch with a voice that could be heard in the neighbouring county. Her beefy, expensive husband looked on gloomily. After much ordering about and many tears, the vault-like doors of the go-anywhere four-wheel-drive vehicle, an essential for Britain's upper-middle classes faced with the infamous and treacherous Himalayan landscape of East Anglia [in truth, all but as flat as a pancake, and as gentle as a labrador retriever napping on a summer evening after supper], were sealed and off they went, unhappily but safely, on their way to an important lunch.

I know which car I would have preferred to be in as a child, but, yes, motoring in the paranoid, safety-conscious, litiginous developed world has long lost its innocence. It is not often fun. It is a duty, if not a chore. This loss of innocence is, I feel, shown throughout the pages of this book. When we selected the pictures, we found ourselves facing a barrier – raised somewhere in the 1980s – when images appeared to become luridly coloured and almost consistently commercial, in an aggressive and humourless sense. In them, no one seems to be enjoying themselves. At the same time, the cars themselves changed considerably, becoming smooth, peardrop-shaped, global, aerodynamic and absurdly "organic', designed to cause the least offence to the greatest number of people worldwide. Or awkwardly, and insincerely, "retro". Gone were national characteristics, idiosyncrasies and innovative design. Regulations in the motor industry are partly the cause of this, yet the more we ploughed through reams of pictures, the less we liked what we saw from the 1980s onwards. This is not entirely due to nostalgia, but

N
1939 MORRIS EIGHT
The quintessence of a "nice drive"; a basic family car that smells of petrol chugging off to the seaside for fishpaste sandwiches and lashings of ginger beer.

O
1962 CHEVROLET IMPALA
In biographies, we rarely hear about the cars famous politicians of the past drove; JFK was US president when this Chevvie cruised Washington avenues. Which power-broker might it have belonged to?

N

because of the way the motor industry, our relationship with the car, styling and photography itself have gone.

I very much like the lack of irony in these old pictures, the sense that the car, for all its mechanical gaucheness and lack of real sophistication, was seen as an adventure on four – and, occasionally, three – wheels. I liked very much looking at snaps of quite ordinary families marking many of the key events of their lives with proudly owned cars in the background. True rites of passage. I liked the idea of a "nice drive", perhaps especially because I was brought up to think, without condescension, that such things were a nonsense. A drive, no matter how brisk and stylish, was a means of getting from one particular place to another, but only at certain times. Long distances were best travelled by train. Cars might ride on low-loader railway wagons on very long journeys, where they might be welcome at the other end, or nosed gently into the pot bellies of Bristol freighters to be airlifted to the Continent. My Uncle Reg, a one-time tea planter from Assam, Eighth Army officer and, in later life, engineer for Vauxhall at Luton, was just about the only person who ever took us for "a drive". Even then, it was brief and designed to show us a new model, Detroit-style Vauxhall or otherwise, to explain the workings of some new engine. My curiosity was insatiable, but I never wanted to go on a "nice drive". Fellow schoolchildren spoke of long, hot, sickly drives in sticky cars reeking of petrol and bound for Pembroke, the Cornish Riviera, the Lakes or even Scotland. They would be stuck in tortuous tail-backs on even more tortuous roads. The Pembroke Coast, Cornish Riviera, Lakes and Royal Scot expresses, meanwhile, roared past proudly on metalled tracks.

In putting the book together, we trawled through our own cupboards and drawers fetching out forgotten family snapshots. It is remarkable just how often cars feature. They are used to mark out our lives. Here we are as babes in arms, then school children on summer holidays, with Mums and Dads, dogs, Uncles and Aunts, and those distant cousins and family friends that, try as you might, you can never put a name to. I see these family photographs abandoned all too often in antique markets and car boot sales and wonder who all these smiling people are staring out at us so optimistically and standing in front of cars. We may not know who the people are, yet we can still recognize the make, model and even the date of the cars. Look at the Ulster family on page 11. I do not know their names, but I can spot, like you, a Rover 2000 hiding behind those natty clothes and smiling faces. And, we can divine from a picture like this some of the aspirations of those who liked to be snapped in front of the chic Rover.

Throughout the book, we have largely selected pictures to show people – whether owners, drivers, film stars or models – with the cars; this is because the car is designed for people and without them it is as lifeless as Tutankhamun's chariots are locked away in the Egyptian Museum in Cairo. Well, not quite lifeless. The best car designs do express a latent energy, so much so that we can't wait to get behind the wheel and zoom off, fueled on the particular dream this or that car promises.

O

A rocket-sled ride in a Monteverdi Hai, a Devon lane bumbling in a Hillman Minx.

What this book tries to show is how cars have become inextricably linked with so many aspects of our lives, including fashion, sex, politics and cinema. This is not a conventional car book, although I think it was meant to start out that way, because although I am fascinated by the history of the automobile and have a great fondness for certain cars, their engineers and designers, I am intrigued by the way they are used as props in our everyday lives. And, equally, how they are so often missing from, say, biographies and studies of the famous. You might imagine from all the millions of words written on Stalin or Che Guevara that they were strangers to the car. But here we see Stalin, the Soviet Union's "Man of Steel" clearly thrilled by his new armour-plated Zis limousine. And how incongruous it is to be reminded, if not here, that Che Guevara, the seemingly ascetic revolutionary hero, drove himself, badly, away from his second wedding in Havana, to fellow soldier Aleida March, soon after the Cuban revolution in 1959; the car was a lurid, bright green, fan-tailed Chevrolet Impala. It rests today in the national motor museum in central Havana. I was shocked, although of course I shouldn't really have been – sometimes a car is just a means of transport, after all – when I first saw this blowsy capitalist plaything. I had imagined that Che would have driven Aleida away in a Jeep. That, of course, is American, too.

This book is a snapshot of these things and not, in any way, a comprehensive account of the social, and much less, the engineering history of the car. The subject is daunting and there are many fine books on individual marques, designers and engineers to keep you researching till the last drop of oil has been drained, profitably, from the deserts of the Middle East, and new types of machines, whether hydrogen-powered, hovering or virtual, or whether donkeys, bicycles and chariots take over when our mad love affair with the car ends. The odd thing is that if this was to happen, it would begin in the developed world. While we might yet gallop along lanes bright with birds, shaded by trees, on super-lightweight chariots, the poorer peoples of the world will be stuck in fuming jams in characterless cars, as they are today in Bangkok and Guangzhou. I look through these pictures, and for all the glorious cars I have driven and owned, the fascinating conservations I have had with engineers and designers from the late, great Wally Hassan – of Bentley and Jaguar fame – to today's young turks at the Royal College of Art in London, I can't help wondering where on earth we've ended up. I envy Tutankhamun's freedom as he rode chariots far more refined than our commonplace cars, with their oily, whirry engines and noxious emissions. Then I remember he died at 19, and that, as yet, we cannot turn the clock back. Even with this thought in mind, I find myself called back to the growl of a Mk2 Jaguar's exhaust, the whine from its Moss 'box, its flickering Smith dials, the deep-growl of Wally Hassan and co's big-hearted straight XK six, the hiss of its SU carbs, the promise of the freedom of the road, in real style and until the next jam.

THE HORSELESS CARRIAGE

After the collapse of the Communist regimes in Eastern Europe and the former Soviet Union, a not uncommon sight in these countries for some years afterwards was that of car bodies being dragged along by farm horses. The "horseless carriage" had got its horses back again.

In a sense, though, the horseless carriage – the name implies a Victorian or Edwardian motor car – survived for many more years than we have been led to believe. It survived in the design of very many car bodies which, up until the Second World War, still retained the look of a horse-drawn carriage. It was not until streamlining came along, and later chassisless construction, that car styling took on a look of its very own. Today, cars may resemble fish, lozenges, eggs and balloons, but their aesthetic is one drawn from their own internal culture, their own DNA as car designers like to say. The horse and the carriage it once pulled have little or nothing to do with the modern car, even if we still refer, charmingly, to "horsepower".

The very first cars were more like ponyless traps than horseless carriages. Have a look at Carl Benz standing beside the single front wheel of one of his original 1885 machines on page 24. It would have gone faster with a pony pulling it rather than a tiny Otto-cycle engine pushing it along, and it does seem to be more a part of the world of reins and oats than cylinders and petrol. Equally, the photograph of Henry Ford at the tiller of his first "quadricycle" (page 25) appears to portray a man in need of a horse.

Of course, early car makers cared for the look of their machines, yet their main concern at the turn of the twentieth century was to make the things go, and go reliably. It took some while before these petrol buggies sprouted roofs, and very much longer before all their many exposed components were pulled together in one smoothly integrated design. Eventually cars became so smooth that they lost such much-liked identifying features as radiator grilles and exposed headlamps – their "mouths" and "eyes" – and both owners and designers missed them. The smooth-as-a-pat-of-butter school of car design peaked out in the 1980s. In the 1990s, "retro" design began to take off, with cars sprouting old-fashioned and often gratuitously antique features: engine-turned aluminium dashboards, fake radiator grilles, and dials that looked like they had been borrowed from a clockmakers at some time during the 1930s.

The dashboard of the Rover 75 was a glorious example of how to turn the clock back the wrong way: a carriage clock on wheels, this bulbous car, sitting awkwardly on wheels a size too small, featured a dashboard littered with grotesque, antique-style instruments. This was not just an embarrassment, but an abuse of a supposed design tradition. Old-fashioned Rovers – and Rovers appeared to be determinedly old-fashioned up until the sleek, motorway age Rover 2000 of 1963 – featured clear, precise unadorned dials in the best British tradition. If Rover wanted to go "retro" in the 1990s, this would have been much more the way to do so.

It had, though, been difficult for many designers to give up the horseless carriage idea. We still talk – well, many of us do – of "horsepower" produced from under the bonnets of cars. The symbol of Ferrari, among the fastest cars of all, is a prancing horse. We have named cars Mustangs and Hunters. We talk of coachwork and of carrozeria as if the makers of Ferrari or Rolls-Royce bodies were engaged with horse and tackle rather than gears and camshafts. We speak of cars getting into their stride as horses do when given a bit of rein. We park cars in garages that look like stables. But then, the car is only little more than a century old, while humans have been riding and working with horses for thousands of years.

Real revolutions in the shape of cars, away from the horseless carriage and towards something new and belonging to the car alone, have been rare. It took ages for designers to see the aerodynamic and structural as well as stylistic advantages of stretching bodywork right over the running gear of cars. Perhaps this did make precious little sense when the car was unreliable or as fickle as a racehorse, and mechanics needed to get to the appropriate misbehaving part without having to struggle with acres of interfering bodywork.

Even then, despite streamlining and chassisless construction, despite so many advances in issues to do with safety, reliability, security, economy, insulation, smoothness and so on, the car has remained, at heart, an old-fashioned machine. For under its bonnet is an Otto cycle engine developed a long time ago that, with all its cogs, gears and whirring bits, is as outdated today as a horse-drawn plough.

And yet, we find it hard to lose a form of traction we have become familiar with and measure the work it performs by the (very) theoretical number of horses it would take to do the same work. While it is true that the internal combustion engine has become ever smoother, more powerful, cleaner, quieter, more reliable and generally more efficient, it is an old-fashioned machine. While aircraft have moved on, where necessary, from the chain-driven piston engines of the Wright Brothers to the turbo-prop and turbo-fan jet, 99 per cent of cars chug on, or race by, powered by petrol and increasingly efficient diesel engines. Car makers are not just conservative for the most part, they know a good thing when they see it.

While there is oil to exploit, plunder and burn, we will rely on methods of propulsion that now seem as old as the horse itself. Eventually, the motor industry will have to move to other forms of power, or we will have to give up our cars or go back to the days of petrol rationing and the revival of micro cars.

None of this will happen just now, of course. In fact, the current trend is for car makers to build ever more powerful engines at a time when roads are becoming increasingly overcrowded and the possibility of driving fast is being reduced day by day. Perhaps this is a final fling before we really do have to think small and green and clean. Or something else altogether. Like the horse, perhaps.

1

1
1885 BENZ
This is one of the world's very first cars, a pretty, lightweight buggy designed by Carl Benz – seen here in 1925 at the controls of the car at the time of its fortieth anniversary with his wife, Bertha, behind the steering wheel. The engine was a single-cylinder 1.6-litre device producing some 3 or 4 hp, allowing the Benz to ride at up to 8mph. Drive was by bicycle chains; these broke on this very first run. Benz, born in Karlsruhe in 1844, built his first four-wheeler in 1893. He shares the honour as the inventor of the car with his fellow countryman Gottlieb Daimler. The companies they created merged in 1926, when Benz was still alive, to form the mighty Daimler-Benz organization.

2

3

2
1886 DAIMLER
Here's the other original car designer and maker, Gottlieb Daimler, born in 1834, being driven by his son, Paul, through the streets of Berlin in the world's first four-wheeler. The car, which looks every inch a "horseless carriage", was powered by a free-revving engine designed and made by Wilhelm Maybach, another formidable figure in the development not just of the car, but of the aero and other engines, too.

3
1896 FORD QUADRICYLE
Henry Ford at the tiller of his very first car. It had no brakes and no reverse gear, but it did have four wheels, an electric warning bell, and it went, more or less. Ford was 32 at the time. His mass-production empire was some years in the distance, but he was on the move, and without horses.

4

5

4
1899 DE DION BOUTON
Ah, monsieurs, which way are we going? The push-me, pull-you look of the 3.5hp Vis à Vis (face to face) De Dion Bouton. Georges Bouton is at the wheel. The science of ergonomics was clearly in its infancy. The car, though, was well engineered and formed the basis of most De Dion cars until 1908.

5
1901 OLDSMOBILE
Production of this handsome, 7hp single-cylinder car ran from 1901 to 1907. It cost $650. The curved dash and slightly, at least, boxed-in bodywork gave the car quite a racy look. It was still easy to imagine a horse pulling it, but the aesthetic of the car was emerging, if slowly.

6
1901 PANHARD ET LAVASSOR
How many moustachioed, boatered and toppered Frenchman can you fit in a vintage car? This is Baron de Zuylen, Comte de Dion, president of the Automobile Club of France, and his contingent at the finish of the 1901 Paris-Berlin automobile race, in his early estate car. You would have needed at least four horses to pull this mighty motorized cart.

6

7
1904 FLINT BUICK
This is Walter L Marr, Buick's chief engineer, and Thomas D Buick, son of the car maker David Dunbar Buick, arriving back in Detroit on a test run from Flint in their prototype car in 1904. Compared to the boys across the road with their bicycles, these pioneering motorists are filthy. The horse looks the other way from the car, as the horse was increasingly to do.

8
1906 STANLEY STEAMER
Locomotives for the road, the extraordinary cars of twin brothers F E and F O Stanley from Newton, Massachussetts, were among the very fastest of their day. They took a while to start up – 30 minutes – but once they got motoring, these comfortable and quiet cars were hard to catch. This is the 1906 model at some speed on an English rally in later years. The 1908 30hp Speedy Roadster was capable of more than 60mph, while in 1906, the Stanleys took the world land speed record in at 127mph in a sensational, low-drag, cigar-shaped racer. Its two-cylinder, 3.1-litre engine was provided with steam by a boiler pressed to a very intense 1,000lb per square inch. Steam, for a while at least, was king on the road as well on the rails and the high seas.

7

2
STANLEY
1906
Stanley
QE
471

9

10

9
1907 BUICK MODEL D
David Buick was born in Scotland. He was taken to America when he was two and eventually became a bathroom plumber. He began making cars in 1902, but sold out to William Durant and the nascent General Motors just six years later. The tough-looking 1907 model featured Buick's first four-cylinder ohv engine. The company established its works racing team that year.

10
1906 BUICK
Ox cart gives way to horseless carriage on a Michigan bridge. These oxen would soon enough be replaced by a Ford pick-up.

11
1909 ROLLS-ROYCE SILVER GHOST
One of the first truly refined cars, the Silver Ghost was a fine machine that could run far, fast, smoothly and reliably. This car is powered by a six-cylinder 7.4-litre engine. Could it fly! Actually, the two severe-looking men on board in bowler hats will be the judge of that. They are Orville and Wilbur Wright, pioneers of powered flight. The Hon C S Rolls himself is their distinguished chauffeur. Rolls died in a flying accident in Bournemouth, Dorset in 1910. His aircraft was a Wright Flyer.

11

12

13

14

12
1911 MERCER 35 RACEABOUT
One of the truly great cars, the lightweight, 70mph Mercer looked right from every angle, dynamic, resolved, all of a piece. The cars won races without special modifications; they were a classic example of what looks right is right (usually). This one is at speed at Santa Monica in 1911.

13
1912 CADILLAC
An important car because it boasted an electric starter. It was this kind of luxury that gradually swayed people with money towards the car. A closed, aluminium-bodied version was available. As the car got more powerful, it could catch up on and finally improve on the comfort of horse-drawn carriages. Powerful engines were needed to carry large and comfortable bodies.

14
1917 MODEL-T FORD
One way of going faster on precious little horsepower… John McLaren and J H Gillmore of the Virginia and Rainy Lake Company out for a ride on the track of the Iron Horse.

15
1921, PORTRAIT OF W O BENTLEY
The modest and deeply talented Walter Owen Bentley (1888–1971) is a true engineering hero. Trained as an apprentice locomotive engineer at Doncaster from 1905 with the Great Northern Railway, he set up his famous company in 1919 after a distinguished war service during which he designed the fine BR1 and BR2 aircraft engines. His mighty Bentley cars were enduring legacies of the days of the horseless carriage, and yet for all their old-fashioned looks – more road-going locomotives than horseless carriages, really – they were gloriously fast and reliable. They won the epic Le Mans 24-hour race five times.

15

16

17

16
1921 PEUGEOT QUADRILETTE
Peugeot's 60km/h baby car brought the horseless carriage to thousands of people who thought they would never be able to own a horse, or even a donkey, much less a car.

17
1926 MODEL-T FORD
Going where no horse can go… a Model-T as a snowmobile. It was the Model-T's ability to outdo the horse even on its own territory – the farmyard, if not the race-track – that encouraged American farmers to give up the horse and turn to cars, pick-ups, tractors and trucks.

18
1920s MODEL-T FORD
Well, here's another fine mess… Stan Laurel and Oliver Hardy caught between two Los Angeles street cars in a sequence from *Hog Wild* (James Parrott, 1930). The Model-T was used and abused in films like no horse ever should or could be. Its comic potential was well established by the 1930s.

18

TYRES
CASTROL

FASTER, FASTER...

"We declare," wrote Filippo Tomassa Marinetti (1876–1944), the iconoclastic Italian poet and polemicist, in his "Futurist Manifesto" of 1909, "that the world's wonder has been enriched by a fresh beauty, the beauty of speed. A racing car with its trunk adorned by great exhaust pipes like snakes with an explosive breath… is more beautiful than the Victory of Samothrace." The winged victory of Samothrace (c190BC) is one of the great classical sculptures. Far from frozen, this Grecian Valkyrie appears to be beating her wings through the air, more beautiful and far faster than any human 2,000 years ago. She was, though, nowhere near fast or beautiful enough for young Italian artists, thrilled by all things modern, in the years leading up to the First World War, when very few cars could top 100mph.

Speed might be beautiful but, like war, it kills. Signs along fast roads make this painfully clear. We know it's true. Yet we know also two other things. Speed is relative. A modern Bristol or Bentley travelling at 100mph or even 200km/h is merely loping along, while a 1.3-litre Morris Marina or Moskvitch 412 at 70mph are at the end of their tether. Both models have their supporters, but only the brave, foolhardy or mechanically insensitive would strive to take them faster than this.

Even then, our speeding British bulldogs are still covering ground at a prodigious rate. Should a tyre burst, a windscreen shatter, a dog saunter into the road ahead, a patch of black ice emerge over the next hill, and these heavy machines may well kill their drivers. Actually very few people have ever been killed in a Bristol and not many more in Bentleys; these are inherently safe cars.

Speed is frowned on by politicians, police (except when they're behind the wheel) and sensible people everywhere. Not surprisingly. Far too many drivers think they have, at the very least, the driving skills of the greatest race and rally drivers, when the truth is that they are probably inept and, as Ralph Nader, the American consumer rights champion, famously said of the rear-engined Chevrolet Corvair, "unsafe at any speed". The testosterone-pumped sales rep hogging your bumper on the outside lane of a motorway when you are making proper progress in heavy traffic in poor weather. The monkey-faced nerd who thinks it's clever to drive a car with a loud exhaust fast through high streets. The chap who swings out of a golf club in a 300bhp coupe, a few gin-and-tonics down the line. The flashing and beeping southern European drivers who, car crammed with children, insist on overtaking uphill on blind mountain bends, and never at any other time, unless a convoy of 40-ton articulated lorries is bearing down, full throttle, in the opposite direction. Speed does indeed kill.

And yet, it has also been one of the key forces at work raising the standard of automotive engineering. A car, like a McLaren F1, that can top 230mph safely simply has to be beautifully and safely engineered. German cars that are expected to cruise serenely along autobahns at 200km/h must, in some ways, be as strong as a Ferdinand-Porsche-designed Tiger tank. Contemporary Formula One racing cars, made of ultra-light metals and super-strong composites, allow a driver to walk out of a high speed crash as if they had hit a bail of hay at 30mph.

Beyond this lies the sheer quest for speed. This begins in many of us at a very young age. We want our scooters to go faster than they can. Then our bicycles. Who can forget the moment they first topped 30mph on two wheels under their own power, or got up the courage to keep their hands off the brake levers as their bike gathered speed inexorably down a long hill with a bend at the bottom?

When you first take to a car, 30mph seems surprisingly fast. By the time you take your test, it seems frustratingly slow. Confidence mounts with occasion and, soon enough, drivers of varying abilities are whizzing along at speeds that were once the preserve of birds of prey for thousands of years, then streamlined express passenger steam locomotives 50 years ago and finally light aircraft today.

Today, German manufacturers, among others, have agreed loosely to a voluntary limit on top speeds; Audis, BMWs and Mercedes-Benzes that could go faster are reined in at 155mph, which is, given the state of most roads known to humankind, quite fast enough for anyone.

There is, then – or so it seems – no need in the future to go any faster than cars can already go. Instead, sporting-minded manufacturers aim for ever quicker acceleration, improved ride, handling and roadholding. And increasingly effective brakes and tyres. All this might seem absurd in many major cities of the world where fast, powerful and heavy cars are used to negotiate journeys at an average speed of less than 10mph.

Speed, though, aside from helping to raise engineering standards, remains a status symbol, of sorts. My car is faster than yours. This, we know, is dumb but, among all too many men at least, this myopic mantra reigns supreme as it has since cars first got above walking pace.

Even so, there is something undeniably thrilling in looking at a photograph of Jim Clark (1936–68) powering a MkI Lotus Cortina around a bend at Brands Hatch, one front wheel pawing the air as the car gallops that bit faster than its designers intended it to. Perhaps, inevitably, speed killed this quietly spoken young farmer, nicknamed "The Flying Scotsman"; his Formula Two Lotus careered off the racing track at Hockenheim in Germany. Car and driver ploughed into trees. They were covering ground, air and fatal ground again, at 170mph.

No fewer than 100 top-rank racing drivers were killed behind the wheel in that fast and dangerous decade, 1958–68. Jackie Stewart, Clark's natural successor and three-time world champion, campaigned effectively for safety on racing circuits. If Grand Prix driving seems less exciting today, it is, aside from the fact that no one ever seems to overtake any more, the result of ever tougher safety measures. They may not seem it – the TV screen belies the fact – but the cars are actually faster than ever.

As for sheer speed, it might take a little time, and even more inclination, for anyone to attempt to top the world land speed record set by Andy Green, an RAF Tornado pilot, in October 1997. Piloting Richard Noble's rocket-like Thrust SSC across the Nevada desert, Green topped 763mph, breaking the sound barrier, for the first time on land, in the process. I wonder what Marinetti, who died in 1944, would have made of that.

19

20

21

19 + 20
1904 FIAT 75HP
Vincenzo Lancia (1881–1937) was a big man in every way. Choosing not to become an accountant, he took up engineering. He was with Fiat in 1899 at the very beginning. Giovanni Agnelli made him Fiat's chief test driver and a member of the Fiat race team. Known as the "Red Devil" for his great speed at the wheel of the blood red Torinese cars, he set lap records but won few races because of mechanical troubles. He did win the 1904 Florio Cup, however, two gruelling laps over a 370km course on dusty roads from Brescia and back via Cremona and Mantua. His average speed was a sensational 115.7km/h (71.88mph). Try that in a modern 75hp Fiat today. Lancia stopped racing in 1906 and founded his own car company with Claudio Fogoli the following year. It is now owned by Fiat.

21
1901 MORS 60HP
The winning cars line up at the end of the 1901 Paris to Berlin race. First over the line, in 15 hours 33 minutes and 6 seconds, was Henri Fournier at the big wheel of a 10-litre 60hp Mors. Second and third place went to the two Panhards on the right. The drive chain on the big Emile Mors racer broke just as Fournier was about to set off on a victory run. Top speeds were up to 75mph at this time.

22
1909 BLITZEN BENZ
Victor Hemery at the helm of the record-breaking four-cylinder Blitzen Benz (Lightning Benz) at Brooklands race track in Surrey. He had been timed at 125.95mph, taking the world speed record from Fred Marriot, who had reached 121.57mph in a Stanley Rocket, a steam car, at Daytona Beach in 1906.

22

23
1909 BLITZEN BENZ
The record-breaking Benz in action again, at Daytona Beach this time, in 1910. The car is mounted on wire wheels and with a polished radiator grille. Barney Oldfield is driving; he took the record at 132.1mph. "You have every sensation of being hurled through space," he said. "The machine is throbbing under you with its cylinders beating a drummer's tattoo, and the air tears past you in a gale. In its maddening dash through the swirling dust, the machine takes on the attributes of a sentient thing… no man can drive faster and live!" He drank and gambled away a fortune, yet died in his bed aged 68 in 1946.

24
1910 BUICK TEN
This is Louis Chevrolet speeding along in Buick Model Ten racer. A muscular, six-footer from Swiss Jura, Chevrolet worked in the bicycle and early French motor industry before emigrating to Montreal in 1900, aged 22. He worked as a chauffeur before turning to racing, where great strength as well as skill was needed to win races at the wheels of these hulking great cars. He won his first race at Morris Park, New York in May 1905 for David Dunbar Buick, a Scottish plumber and inventor turned car maker in 1899. Chevrolet's brothers, Arthur and Gaston, were also a part of the team. Louis would eventually become famous for the cars that still bear his distinctly racy name.

25
1910 BUICK TEN
Here's "Wild" Bob Burman ready for a race on Daytona Beach against a Wright aircraft during the 1910 Carnival of Speed. A nail-biter, eh? Cars and trains could easily outrun early aircraft and continued to do so for some years to come.

23

24

25

26
1913 PRINCE HENRY VAUXHALL
Laurence Pomeroy trained as a locomotive engineer before joining Vauxhall in 1905. In 1910 he entered Prince Heinrich von Preussen's "tours" – a high-speed German rally – with a three-litre model. This led to the four-litre Prince Henry Model. With 75hp, it was smooth, if not quiet, and capable of a very relaxed 60mph cruise and a top speed of 75mph. For its period, it was a very accomplished machine, and a big success in trials and rallies.

27
1921 VAUXHALL 30/98
A 4.5-litre E-Type Vauxhall at full tilt, throwing up dust at a speed trial. These events were immensely popular between the two world wars, allowing amateur drivers to try their luck and risk their necks. The Vauxhall, designed by Laurence Pomeroy (1883–1941), who trained, like W O Bentley, as a locomotive engineering apprentice, was one of the finest sporting cars of its time, even though it was based closely on the Prince Henry Vauxhall, dating from 1913.

28
1932 MG J2 MIDGET
Not exactly quick in absolute terms, but there was immense fun to be had seeing how quickly you could get your new sports car up a muddy hill without sliding off or getting stuck. This is a delightfully English scene at the Abingdon trials, Oxfordshire, in 1932. The chaps squeezed into the tiny cockpit wear their caps at jaunty angles while the bloke in a cap sitting on the ragged stone wall gives them a disapproving look; call that a car, you nonces. In fact, the J2 (1932–34) was a fine little machine. Although its 847cc engine was only good for 65mph or so, what it did, it did in great style and was a delight to drive. It gave many people their first opportunity of trying out a real sports car.

26

27

28

29

29
1924 BUGATTI
Not to worry, old chap, the jolly old wheel's gone and dropped off. This is Raymond Mays, the renowned English racing driver, looking wonderfully unconcerned as he loses a wheel at speed from his Bugatti. Ettore Bugatti, the great car designer and maker, had been so impressed by Mays's trouncing of land speed record-holder Malcolm Campbell on the Shelsey hill climb, that he gave Mays one of his cars for free. Some gift. The unflappable Mays went on to help create two legendary British racing marques, ERA (English Racing Automobiles) and BRM (British Racing Automobiles).

30
1915 DUESENBERG
Bill Chandler at speed in one of the great American cars on the new Des Moines Speedway, Iowa, on August 7, 1915. The cars lapped at more than 90mph – the track lived up to its name.

31

32

31
1937 ROLLS-ROYCE PHANTOM III
A magnificent Park-Ward-bodied 7.3-litre V12 Phantom III sets the pace at Aintree, in the wet, in 1937. The car boasted independent front suspension adapted from a General Motors system. Built from 1936 to 1939, it was comfortable, drove well and, for such a big car, was quite quick enough. In October 1936, *Autocar* recorded a top speed of 92mph, 0–60 in 16.8 seconds… and an overall fuel consumption of just 10mpg.

32
1935 TATRA 77A
Austrian engineer Hans Ledwinka's streamlined masterpiece, the exquisitely engineered rear-engined, air-cooled six-seater V8 Tatra was revealed to an astounded public in 1934. Ledwinka had teamed up with fellow engineer Erich Uberlacker and Hungarian aerodynamacist Paul Jaray to shape this revolutionary 140km/h machine. The 1935 model boasted many modifications, including a larger 75hp 3.4-litre, and a top speed of 150km/h. The Ledwinka streamliners were much prized by German army officers after the annexation of Czechoslovakia by Nazi Germany in 1938. A number were killed driving too quickly; the Tatra was fast in a straight line, but a handful around corners. The army banned its use.

33
1939 LAGONDA V12
Here is one of Britain's finest pre-war cars, the Lagonda V12, engineered by W O Bentley, in racing guise at Le Mans, 1939. With its lightweight aluminium body and extensive use of alloys, the great 230hp 4.5-litre car was fast. Almost unmodified from the stock 100mph road model, the two Lagondas (you can just see the nose of the second one here) entered for the 24-hour race came in third and fourth. The cars rode well and were equipped with all-synchromesh gearboxes – rare before World War Two – and powerful hydraulic brakes. They remain a joy to drive more than 60 years on.

33

34

34
1937 MERCEDES-BENZ W125
These mighty Mercedes were fully capable of 200mph. Designed by the young Rudolf Uhlenhaut, who could all but match the German racing team's top drivers, the car was equipped with a 646hp twin-overhead-cam, four-valves-per-cylinder, supercharged 5.7-litre V8. Here is one the cars taking off in the 1937 British Grand Prix at Donington. This race was won by Bernd Rosemeyer in a rival C-Type Auto-Union.

35
1950 ALFA ROMEO TIPO 158
Whether the cars are going forwards or backwards – as it appears here – the Monaco Grand Prix remains of the one of the highlights of the racing year. The tight circuit weaves through the streets of the tiny principality of Monte Carlo, the cars driven at seemingly impossible speed. This is Juan Manuel Fangio, generally considered to be the greatest racing driver of all, on his way to winning the first Monaco Grand Prix in 1950. That year, "Alfetta" drivers – the "little" supercharged 1.5-litre Alfas, timed at up to 192mph elsewhere, were essentially pre-war designs updated for the early post-war international races. At the end of the 1950 season, the Alfa team came home first (Farina), second (Fangio) and third (Fagioli).

36

36
1954 MERCEDES-BENZ W196
A poster celebrating the 1-2 win of Fangio and Kling at the 1954 French Grand Prix. This was Mercedes-Benz's triumphal return to GP racing, the first time in fact since 1939. Alfred Neubauer, 1930s' team manager, was back in place. His cars were the beautiful W196 streamliners, their 257hp 2.5-litre eights screaming up to 8,500rpm. These are among the very best looking of all racing cars, and were very effective, too.

37
1956 LE MANS RACERS
Rain did little to dampen the enthusiasm of the 250,000-strong crowd as they cheered on the entrants of the 1956 Le Mans 24-hour race. At that time, the drivers had to sprint to their cars before roaring off towards the famous three-mile Mulsanne straight and attaining mercurial speeds. An Ecurie-Ecosse D-Type Jaguar won the race, with a works Aston Martin DB3S and Ferrari 625LM in second and third places. The winning average was 168.122km/h, with the fastest lap at 186.383km/h set by Mike Hawthorn (sixth place) in a D-Type Jaguar. Le Mans remains one of the most charismatic and best loved car races.

OVERLEAF
38
1949 BRM V16
Breaking the sound barrier. Veteran Bugatti, ERA and BRM driver Raymond Mays negotiates the great British workman – on bicycles – and parked Vauxhall Victor, VW Beetle, BMC 1100, among others – on the occasion of a gathering of the Club International des Ancien Pilotes, July 17, 1967. Actually, this is probably a later model, because the cars were made and raced up until 1954. Although powered by ingenious 1.5-litre V16s, the BRMs proved unreliable and won few races. The sound they made was unforgettable, though.

39
1969 GP RACERS
Cars scream past photographers and spectators sitting and standing right beside the track at the 1969 French Grand Prix. This seems all but impossible from the safety-conscious vantage point of 2003. For the record, the race was won by Jackie Stewart in a Matra-Ford at 157.251km/h.

37

SOLEX
SOLEXINE
KLÉBER-COLOMBES
NEWTON
BHB
KLG
DUNLOP
FERODO
POLAROIL
Huiles IGOL
LUCAS
A
B
C
D
17
16
15
14
12
11
10
9

38

39

40

41

40
1965 LOTUS-CORTINA MK1
Jim Clark, the "Flying Scotsman", lifts a wheel of his white and green-striped Lotus-Cortina at Brands Hatch in 1965. Clark was among the all-time racing greats, and the souped-up, twin-cam Cortina one of the delightful surprises of 1960s motor racing. It seemed as if anyone could have a go. But, even if they could afford the Lotus-developed Ford, would they ever get to drive this well?

41
2003 CHIP GANASSI RACING HAVOLINE DODGE
US muscle cars zooming around the California Speedway at Fontana, CA. The occasion is the NASCAR Winston Cup Auto Club 500 race on April 27, 2003 and Jamie McMurray leads the pack in a Dodge. The cars, like most contemporary racers, are plastered in corporate advertising. They can lap at more than 180mph.

42
2003 FERRARI
This is Michael Schumacher, the great champion German racing driver of the past decade, refuelling during the 2003 Malaysian Grand Prix at Kuala Lumpur. By this time, not only had GP racing, originated in France, gone global, but the cars could be serviced in a matter of just a few seconds at pit stops. This needs to be seen to be believed. The smartly uniformed mechanics resemble worker bees droning around their Queen. Or King, in this case, although Schumacher came sixth in race, yielding first place to the Finn, Kimi Raikkonen driving a McLaren and lapping at an average of 201.629km/h.

42

43

2003 NASCAR WINSTON CUP VIRGINIA 500

Like a giant Scalextric slot-car circuit… but these are real muscle cars bunched together at great speed. Crowds worldwide never seem to tire of watching brightly coloured cars going round and round. They have done so now for a century. This is a scene from the Martinsville Speedway, Virginia, April 13, 2003.

43

Series
NASCAR FOX
8
25
24

EVERYMAN

When did the car become popular? When ordinary people could afford it. The earliest cars were craft-built designs and so, almost by definition, the plaything of the wealthy. It was, of course, Henry Ford who set out, in his words, to "build a motorcar for the great multitude". Ford's dream was of a mass-produced automobile, an automotive fanfare for the common man; in his mind, it was to be a staple commodity "just like one pin is like another pin when it comes from the pin factory, or one match is like another match when it comes from the match factory".

Ford built his first car in 1896. Ten years later this steely, anti-semitic eccentric set up a secret drawing office in his Detroit factory and began work on his famous Model-T. The result was a light, strong, easy-to-drive machine powered by a 2.9-litre, 20hp four-cylinder engine. It went on sale on October 1, 1908, price $825.

This was not particularly cheap, and although 10,000 Model-Ts were sold in the first year of production – an industry record – Ford wasn't satisfied. The cars were still built using traditional methods. What was needed was true mass-production. "I'm going to democratize the automobile," he said in 1909. "When I'm through, everybody will be able to afford one, and about everybody will have one."

A year later, Ford opened his revolutionary Highland Park factory, which was designed by the prodigious and brilliant architect Albert Kahn, who later went on to build hundreds of factories for Stalin in the Soviet Union. Ford, although a fervent anti-Communist and union-breaker, was, in his own primitive capitalist way, a man of the people, too. Production of the Tin Lizzie, the sole Ford model for many years, rose from 19,000 in 1910 to more than 78,000 four years later. And the price of the car began to fall. In 1912 it was $575 – less, for the first time in the history of the motor car, than the average annual wage of a working man. The Model-T's market share, 9.4 per cent in 1908, was a staggering 48 per cent at the outbreak of the First World War in Europe.

New assembly lines based on meat processing factories in Chicago boosted productivity. Ford's endless chain-driven, waist-high, conveyor belt lines, which allowed workers to stay put in one place on the factory floor, were in place by spring 1914. Reduced to the level of human robots, workers needed fewer and fewer craft skills and were paid correspondingly less than in rival car plants. Ford said his methods gave employment to ordinary labourers, but staff turnover was high, and the mechanistic methods of his non-stop, clockwork, all-but-inhumane factories were ultimately to be decried in such films as Fritz Lang's *Metropolis* (1927) and Charlie Chaplin's *Modern Times* (1936). In 1932, Aldous Huxley published *Brave New World*, a novel set in the future when the philosophy of what he called "Fordism" had reduced working people to automata.

Ford's introduction of a $5 eight-hour day plus profit-sharing for workers in 1914 seemed a radical, even socialist move, yet it was a canny ploy to reduce the turnover of his workforce. By 1921, the Model-T held 60 per cent of the new car market, and a titanic new factory for 75,000 employees was built at River Rouge outside Detroit. Here a Model-T could be conjured from raw materials to finished product in just 41 hours. In 1924, the ten millionth "Flivver" rolled off the production line, by which time the car's price had fallen to $295. Sales finally fell as smart, cheap and modern new Chevrolets reached the market. Model-T production ended on May 25, 1927, and no fewer than 16 million had been made – a record finally broken by Ferdinand Porsche's Volkswagen ("People's Car") Beetle, which remains in production in Mexico and Brazil in 2003.

This was not exactly the end of the road for the Tin Lizzie. Although superseded by Edsel Ford's successful Model-A, the Model-T was seen everywhere, not just across the United States but the world; wherever, in fact, a car could make tracks. From co-starring roles in Laurel and Hardy comedy "shorts" to its workaday role as farmer's pick-up, the car was a symbol of the United States as the country began its ascent into the role of international superpower and today's formidable empire. The Tin Lizzie ought, perhaps, to feature, alongside George Washington and the Bald Eagle on the US Treasury's universally recognized dollar bill. Will Rogers, the cowboy philosopher, humorist and Ziegfried Follies star, had once nominated Ford for president, saying "there's no reason why there shouldn't be a Ford in the White House; they're everywhere else".

The first Fords, the original 8hp Model-A, had gone on sale in Europe in 1904, but Ford really only took off in the Old World with the launch of the Model-T at Olympia, London in November 1908. Three years later Ford opened its first European assembly plant at a former tramcar works at Trafford Park, Manchester. Production later moved to Dagenham on the Thames marshes, where the Depression era 933cc Ford Eight was built from 1932. In 1935, this became the famous £100 Ford Popular, the cheapest factory-produced British car ever, and the first of a long line of basic family Fords that lasted until the arrival of the highly styled 105E Anglia launched the same year as BMC's radical Mini in 1959.

Even then, Ford – even when making the Le Mans winning GT40s and Executive saloons – stuck to its blue collar roots with that long run of Cortinas, low-spec Grenadas and cheap and cheerful Fiestas.

Ford had set the scene for commonplace cars. Before and after the demise of the epoch-making Model-T, manufacturers across Europe worked up designs for every conceivable configuration of small, popular car, from the Volkswagen, conceived in Nazi Germany, through baby Austins, the tin-shed-style Citroën 2CV, tiny minibuses from Fiat, a cadre of Communist era cars – rear-engined, two-stroke – from East Europe and the Soviet Union, to the mini cars celebrated elsewhere in this book. Whatever form they took – good, bad or indifferent – they opened up the world of motoring to everyman, and woman, and, however tortuously, their DNA was encoded in Henry Ford's multi-million-selling Model-T. Yet, as Will Rogers said of Henry Ford,"it will take a hundred years to tell whether he helped us or hurt us, but he certainly didn't leave us where he found us."

44

45

46

44
1924 MORRIS COWLEY
Standing at the side of a damp London street when fairly new, this four-seater Morris was a sturdy, mass-produced product from William Morris's Oxford factory. Morris, born in Worcestershire in 1877, started out as a bicycle maker, like so many early car manufacturers. He built his first car, the Oxford, in 1913. The Cowley dates from 1915 and was progressively developed over the next 15 years. Its distinctive rounded radiator encouraged its nickname, the Bullnose Morris.

45
CITROEN B14, 1926
André Citroën (1878-1935) preferred a playboy lifestyle to tinkering with engines, like most of the pioneers of motoring. However, he possessed, aside from good taste and a fine intellect, a flair for publicity and for shaping popular taste. His first small car, the 856cc Type C Cloverleaf of 1922, might have been cheap, yet buyers could specify a glamorous torpedo-shaped body, while, in defiance of Henry Ford whose methods he adopted, the cars were only available in canary yellow. The B14, a handsome 1.6-litre four-seater, saluted here by an impeccably dressed monsieur, was an early steel-bodied design.

46
AUSTIN SEVEN, 1922
Here are some 1920s flappers up for a lark in and astride their baby Austin. The "Chummy" – it certainly was – was designed in Sir Herbert Austin's billiard room at Lickey Grange, his country house. The principal draughtsman was 18-year-old Stanley Edge. He took his inspiration from the Peugeot Quadrilette and Bebe models. Priced at £165, the 696cc open-top car sold slowly at first but, with a 747cc engine from 1924 and many other modifications and body styles over the years, it became popular with both the wealthy as a runaround (it was the Mini of its day) and the ordinary families for whom it was conceived. It was replaced by the Big Seven in 1937.

47
1935 CHEVROLET SUBURBAN
Family, dog, servants – the revolutionary Suburban could carry the lot. This was the very first SUV, a genre so popular with American families and gangsta-rappers today. The Suburban featured a steel body on a half-ton truck chassis, three rows of seats for eight people, a tailgate and strong performance from its 90hp motor. One of its successors carried me safely from Amman and across the length and breadth of Saddam Hussein's Iraq in 2002. I have reason to be grateful to this American automotive legend.

48
1940 OLDSMOBILE STATION WAGON
When this house on wheels was launched, station wagons accounted for just 1 per cent of the US car market. Manufacturers and buyers alike thought of wagons as commercial vehicles. This Oldsmobile marked a sea change in the type's fortune. Not only was it practical, but it had a lively performance from its six-cylinder, 230 cubic inch, 95hp engine and was made easy to drive with the option of Hydra-Matic (automatic) transmission.

49
NANTUCKET BEACH, JULY 4, 1925
As many cars as people. This is the famous Massachusetts resort in the days before parking restrictions, and at this date, such a scene could really be found only in the United States; the car had taken off like hot cakes.

47

48

49

50

51

51
1959 DKW UNIVERSAL
A happy post-war German family camp with their rare, semi-streamlined, three-cylinder, 900cc estate car. Throughout the 1950s, European manufacturers worked up dozens of ingenious economical family cars. DKW was originally founded by a Dane, Skafte Rasmussen, in 1916 to make steam cars (hence the name, Dampf Kraft Wagen). The first production cars, using petrol engines, were built in Chemnitz from 1928. DKW was absorbed by Auto-Union in 1932. Refounded in the 1950s, it ceased production of its charming cars, including a miniature Ford Thunderbird lookalike, in 1968.

50
1956 FIAT MULTIPLA
Used as Roman taxis, family runarounds and commercial vehicles, the Fiat Multipla was a cleverly engineered miniature minibus, or MPV (multi-purpose vehicle). Designed, off the back of the baby Fiat 600, to carry six people – in practice many more – it was ideal for the sizeable Italian family of the 1950s. Ingeniously packaged to maximize interior space, it was also great fun to drive.

52

52
5 A.M., SEPTEMBER 3, 1967.
STOCKHOLM
Here are the Swedes changing over from driving on the left to driving on the right. Quite why it has been so important for the vast majority of the world's nations to drive on the right remains a mystery, especially to the British and Japanese. Does it really make any difference?

53
1935 VOLVO PV36 "CARIOCA"
Only 500 of these striking and technically advanced streamlined Volvos, designed by Ivan Orberg, were made between 1935 and 1938. But here is proof that the makers of one of the world's most endearing and enduring people carriers, the Volvo, have been able to produce glamorous as well as universally acclaimed workaday cars. Volvo is Latin for "I roll"; the PV36 rocks, too.

54
NEW YEAR'S EVE, 1974.
ST GERMAIN.
CITROEN 2CV.
Mechanical genius, and a symbol of France for 40 years, the 2CV was one of the most remarkable and original cars of all. Designed by the brilliant André Lefebvre – engineer of the Citroën Traction Avant and, later, the DS – the original brief for the TPV (Toute Petite Vehicle) was for a lightweight car capable of driving across a field with a basket of eggs without breaking a single one. Developed from 1938 to 1948, and sold at first in just one colour (grey), it was a huge and enduring success. One of the truly great cars.

53

54

55

56

55
1961 MORRIS MINOR
Here's the one millionth Morris Minor being inspected at the end of the Cowley production line. Designed by Alec Issigonis, of later Mini fame, the much-loved jelly-mould-shaped car was produced from 1948 to 1971. It remains very much a part of the British roadscape more than 30 years later.

56
1956 RENAULT DAUPHINE
Palatial thoughts after a drive in a true people's car. The Dauphine, the automotive equivalent of a goldfish – have you looked one in the gills, I mean grille (or lack of one) – was produced from 1956 to 1968. Powered by a 32hp 845cc engine mounted at the back, it was a spritely performer, and a very proper four-door saloon. It was originally to have been called the Corvette, but for obvious copyright reasons, the name was changed; and it was never that fast, even when tweaked for racing by Gordini.

57
1962 SUNBEAM RAPIER MK3A
A handsome car from Rootes that did well in rallies, the Rapier was a sporting saloon, made from 1955 to 1967, from a company best known for mass-producing American-style post-war saloons. This one boasts disc brakes up front, a free-revving 1596cc engine and side windows that disappear into the bodywork. The Rapier marked the beginning of the end of austerity economics in Britain; motoring could be fun again.

57

58
1948 TATRAPLAN
This is one of the smaller of Tatra's glorious, streamlined, fin-tailed, rear-engined saloons. The original sci-fi-style Tatras were designed by the inventive engineer Hans Ledwinka in the mid-1930s. Ledwinka had also built prototypes of a people's car similar in concept and looks to the Volkswagen before Porsche unveiled his people's car some years later. The economy model T600 Tatraplan, engineered by Julius Mackerle, was powered not by a mighty, air-cooled V8 as the most famous of these legendary Czech models, but by a 50hp, 1.95-litre four. It was smooth, stable and very quiet.

59
1956 WARTBURG ESTATE
During the Russian invasion of Nazi Germany, the BMW factory at Eisenach was taken over by the Soviets. Post-war, it continued to make BMWs under the name EMW. Finally, in 1956, the old models were replaced by a new marque: Wartburg. The cars were quite big and practical but relied on smoky, two-stroke engines. They were always a bit of a laughing stock in the West, bought by left-leaning families fond of camping, sandals and suspicious politics. This 1956 estate was available with either a 37hp 900cc or 50hp 992cc motor.

60
1965 MOSKVITCH 412
Production at the newly designated MZMA (Small Car Moscow Factory) began after the Great Patriotic War (1941–45) with a pre-war Opel Kadett copy dubbed the Moskvitch, or Muscovite. The 412 shown here was the staple Moskvitch of the 1960s and into the 1970s. Simple, crude, rugged and very easy to maintain, the 1.5-litre car won few friends in the world beyond Brezhnev's; but they were much valued in the former Soviet Union. Here, an heroic young proletarian beauty has bought her Moskvitch some lovely flowers.

58

59

60

61

61
1937 SKODA POPULAR
Skoda was long considered a joke outside the former Communist Block until its revival by Volkswagen in the 1990s. But, in fact, the Czech company has a long and distinguished history. The 1937 995cc Liduska, or Popular, model was a fine and handsome little car; it adopted contemporary American looks in 1939.

62
1970 SKODA 110R
The 1970s was a tasteful decade, East or West, as you can see in this picture. The car, a fastback, sporting, rear-engined, Czech model, is actually remarkably good fun to drive; a kind of Porsche 911 for everyman. The girl in the picture certainly looks impressed. The five-speed car boasted 62bhp and a top speed of 140km/h. The 110R was successfully raced and rallied. The most powerful version – there was just the one – had an alarming 697bhp, a real sting in the tail, and recorded 348km/h on an autobahn outside Berlin.

62

63

63
1966 WARTBURG KNIGHT
Here's the new 1966 353 model built at the former BMW works at Eisenach. The body was light and bright, but the mechanics were based on a separate chassis. The engine, as before, was a sturdy, three-cylinder, one-litre two-stroke. Still, our fashionable *frauleins* seem happy with their new people's car. Production of this model ceased in 1976, and of all Wartburgs in 1991.

64
1973 FORD CORTINA MKIII
Well flash, my son, but what are you doing visiting some crumbling old half-timbered pile when you could be flash-harrying down to the yacht club at Burnham-on-Crouch in your brand new – and most probably bronze – 2000E Cortina? This loud, brash Cortina made its debut in 1970. It was very '69 Dertroit, with its Coke bottle styling, acres of chrome, recessed instruments and generally aggressive stance. It was good to drive, simple, reliable and sold like pints of lager in an Essex pub. Since the Second World War, Britain has straddled US and European culture; in the case of the MkIII Cortina, it was Uncle Sam all the way, although without V8 muscle.

64

THE NATIONAL TRUST
INFORMATION ROOM
ENTRANCE TO THE
GUILDHALL
OPEN
Parts of the Guildhall
may be closed for community use
FORD
AOO 955M

65

66

65
1965 FIAT 850 COUPE
A smart, miniature, mass-market "GT". This is the Fiat 850 of 1964 rebodied stylishly by Centro Stile Fiat – the company's in-house design department – and given some extra oomph with a 47bhp, rather than a 37bhp, 843cc rear-mounted engine. A delight to drive and capable of 140km/h, this was Fiat at its 1960s best.

66
1968 FORD ESCORT TWIN-CAM
Ford could produce some very fine sporting cars when it set its corporate mind to it. The Ford Escort MkI was the first car from Ford Europe, a merger of Ford UK and Ford Germany. The highly successful sporting version shown here had a 110bhp Lotus-Ford 1,558cc twin-cam four squeezed tightly under its bonnet. The car was good for 115mph and behaved extremely well. Just 1,200 were made before production ceased in 1970. The Escort itself went on for years to become one of Britain's most popular cars.

67

68

67
1960 RELIANT REGAL MKVI
Look, we can bury it over there, have our picnic and go back up to town by train! Gosh, can we, mummy? I don't want to be sick over my new surfboard again. The first of these bubbly three-wheelers went on sale in 1952. They were improved several times during the 1950s, although all retained the tiny, side-valve, 747cc four-cylinder engine that had first powered Austin's baby Seven in the 1920s. Well equipped and well made, the Regal was regarded fondly, although this family would surely soon be dreaming of a new BMC 1100 or Ford Cortina. Reliant, founded in 1935, went on to build many more three-wheelers, but it also produced the conceptually brilliant Scimitar GTE from 1968, a 120mph glass-fibre GT estate that won it many new friends, including Princess Margaret. Regal indeed.

68
1965 MG MIDGET
Often advertised with clueless dolly birds clutching the gear stick or handbrake, the Midget was a classless, affordable sports car. It was at its prettiest here in the mid-1960s.

69

69
1968 MERCEDES-BENZ 300SEL 6.3
One horse power meets 250hp in the guise of the 300SEL 6.3 saloon favoured by racing drivers, rock stars and discreet business executives. The clever thing about this genuinely fast V8 Mercedes is that it was, and remains, so gloriously discreet. Remove the badges at the back and it could be a much humbler machine altogether. Mercedes had a genius for building cars that might serve – and for years – as taxis, family hacks or the intercontinental mounts of professional racing drivers, and yet all look much the same, except under the bonnet. One of its favourite advertising slogans of the 1960s was "a car your son will inherit". Add "daughter" today, and not too much has changed, although Mercedes saloons have lost the almost commonplace look they once had that made them, almost if not quite, a part of the crowd.

70
1985 FORD ESCORT USA
C'mon, sugar, please come for a ride in my Ford Escort. Gee, Brad, I'm sorry, I just met a guy with a Bond Minicar … Enough said.

70

NURSERY SERVICE STATION
SALES OFFICE
Lucas Motostockist
NO ENTRY
500
STAMPS
IF YOU JOIN
THE AA HERE
SUPER
MULTI-
GRADE
45
PNO 529R

71

72

73

71
1978 FORD CAPRI 1.3
Go on, Stan, push it harder, will yer? And, Kevin, you can stop being sick in the back before I give you a good slap – have you heard me? Bleugghhh… sorry, Mum. A bottom-of-the-range 1,296cc Capri getting a push-start (or is he just tanking it up with four-star?) in a smart English service station. The Capri, first launched in 1969, was a Mustang "personal coupe" for the British and German markets. It had a reputation for being a mount for wide-boys but, in fact, it catered for a wide range of tastes and incomes. A 1.3 like this MkII model (1974–78) was positively mild mannered compared to the 3.0- and 3.3-litre beasts that were available. Production of the Mk III Capri ended in 1986. Although named after the Italian island, the car was always called a "Ca-pree" rather than a "Cap-ri"in Britain. It was that kind of motor, mate. So don't come over all fancy on me.

72
1969 DODGE CHARGER
"By the time we got to Woodstock/ We were half a million strong… We are stardust /We are golden/And we've got to get ourselves /Back to the garden…" sang Joni Mitchell, although she didn't add "…and this Dodge Charger out of the mud, man". Woodstock was the biggest pop festival yet. It wasn't quite as beautiful as it was cut out to be, as you can see from this snap of youngsters trying to get their Dodge back on the road. Hey, how much money did these Love Generation, no-possession kids have? That Charger's brand new, man.

73
1999 HINDUSTAN AMBASSADOR
This Calcutta street scene would have changed little since the Hindustan was first made near Calcutta in 1948. At one time in the 1980s, virtually every new car built in India was this 1948 Morris Oxford modified slowly over the years for Indian driving conditions and the ways of crowded Indian cities. This example is adorned for a wedding. The cars, India's Model-T Ford, are very much still in production, although in recent years the venerable BMC B-series engine has been replaced by a smoother, 75bhp, 1.8-litre Isuzu unit. Ambassadors are now seen on the streets of central London in the guise of Karma Kabs, adorned with flowers and smelling of incense and petrol.

Austin Seven
183 WHK

POWER AND POLITICS

During a hugely popular public tour of the German province Thuringia in 1934, Adolf Hitler turned to his motoring companion, the architect Albert Speer, and said, "Until today, only one German has been hailed everywhere like this: Luther. When he rode through the country, people gathered far and wide to cheer him, as they do for me today". Architect and Führer were travelling in an open-top Mercedes-Benz strewn with flowers thrown at them by well-wishers lining village streets and country lanes.

Hitler did not drive, but along with power, he loved speed and cars, and most of all the succession of mighty open-topped Mercedes-Benzes he owned as Leader of the German people. He was, says the German author, Wulf Schwarzwaller, "not only the company's best customer, but he was also its best advertiser". So closely was Hitler associated with the Daimler-Benz company that the three-pointed star that adorned the massive radiators of Mercedes-Benz cars was, during the Nazi era, all but synonymous with the swastika. Hitler, who, of course, nurtured the KdF-wagen ("strength through joy" car, or Volkswagen) was close enough to Mercedes-Benz to advise on the design of its cars. "I can claim credit for the things that make the Mercedes cars so beautiful today," he said. "In drawings and designs, I tried hard, year after year, to perfect that shape to the utmost."

That shape was the iconic form of the massive machines from Mannheim that culminated in the Type 770 Grosser model of 1940. One of these five-ton, 230hp cars was delivered new to Berlin by Mercedes-Benz for Hitler's triumphal parade through the city on July 19, 1940 after the fall of Paris. The car, unlike Hitler, survived the war and is now on display at the Canadian War Museum, Ottowa.

Hitler's famous morning tour of Paris, with sculptor Arno Brecker and Speer, on June 23, had been in a muscular, three-axle Mercedes-Benz G-4 cross-country tourer.

Other Nazi leaders were quick to emulate their car-mad boss. Heinrich Himmler liked to tour his concentration camps in a Mercedes cabriolet; its registration plate was SS-1. Hermann Göring opened the new autobahn from Berlin on April 4, 1936 by driving its length, at speed, in his truly beautiful eggshell blue special 540K Special Roadster. His passenger was Dr Fritz Todt (1891–1942), Hitler's General Inspector of Road Building and creator of Germany's autobahn network. "I've had some queer drivers in my time," said Hitler, recalling the good old days, 1942. "Göring made a point of always driving on the left-hand side of the road. In moments of danger, he used to blow his horn. His confidence was unfailing, but it was of a somewhat mystic nature!"

Hitler understood the value of the car better than any other politician before him, and most since. The car enabled him to move quickly around the new Reich, to be seen and greeted by the German people. The cars he chose were also impressive, imperious machines that conveyed a proper sense of might and awe. He knew that he could easily be assassinated standing and riding in open cars, but this was a part of his image and myth. The Führer was as indestructible as his Grosser Mercedes. An armour-plated saloon was made for his 50th birthday in 1939. This was a Type 770K W150 II. It weighed more than five tons but, powered by a 394hp supercharged eight-cylinder engine, it could top 112mph. Hitler, however, rarely used it. The point of his car journeys was to be seen, not to be hidden away.

No doubt John F Kennedy felt much the same way when he was driven through Dallas at midday on November 22, 1963. In fact the US president had insisted personally that the bubble-top roof of the 1961 Lincoln Continental should be removed if the weather was fine. That decision cost him his life.

The car itself had been well chosen. Much as Hitler's Grosser Mercedes reflected his sense of power, his style of politics, so the smooth, chrome and fin-free '61 Lincoln matched and mirrored the youthful President's decidedly modern style. His was the age of jet airliners, nuclear bombs, IBM electric typewriters, computers, *Breakfast at Tiffany's*, "The Dick Van Dyke Show", "Mr Ed" and "Car 54, where are you?" And of his wife's impeccable and much copied dress sense. His apparently glamorous administration was spoken of as a new Camelot. His charger was that elegant Lincoln, the sort of car forever associated not just with Kennedy's death, but also with adverts in *New Yorker* magazine, dry Martinis and Park Avenue lifestyles. It seemed exactly right.

Just as Tony Blair's Ford Galaxy "people carrier" was exactly the right choice for a young Prime Minister clearly in love with family, America and populist politics. Or as Mrs Thatcher's penchant for Jaguars was a neat reflection of her sort of Britishness: forward looking, dynamic, yet, by jove, underpinned by old-fashioned values.

The car served as a prop, or "extra", in many of the great political dramas of the twentieth century. When the Austro-Hungarian Archduke Franz Ferdinand and his wife, the Archduchess Sophie, were assassinated by Gavrilo Princip on June 28, 1914 outside Schiller's store on Sarajevo's Franz Joseph Street, the first shot of the First World War had, effectively, been fired. The Archduke, a brave chap, was, like JFK half a century later, travelling in an open-topper. This 1910 model belonged to Count Franz Graf Harrach. Appropriately, perhaps, it was painted blood red. Some 8.5 million soldiers died in the First World War, and God only knows how many civilians; remarkably, the car that drove Franz Ferdinand and Sophie to their untimely deaths survives in the Museum of the History of the Army in Vienna.

World wars led to the development of specialist vehicles, the Rolls-Royce armoured cars of the First World War, Willy's Jeep and the Kubelwagen, Ferdinand Porsche's military version of the Volkswagen. Today, the Jeep, a car that in a strange way is connected back to Franz Ferdinand's assassination (would Hitler have been elected Chancellor, would there have been a Second World War, if Germany had not been so aggrieved by the reparations imposed on it in 1918?), has been transformed into a general purpose family runaround. What were once military vehicles, steeped in political blood, and battlefield mud, are now machines for running children to school. This may seem absurd, but better gross SUVs (special utility vehicles) and peace than Grosser Mercedes and war.

74

75

74

1904 AMERICAN MERCEDES

Native American warriors seduced by white man's technology? Or Geronimo and friends out for a joy ride in a captured automobile? This is a puzzling picture. Geronimo, an Apache, was born in 1829 in what later became New Mexico. He was the last of the Indian leaders to fight a war against US troops. Captured in 1886, he was a prisoner of war until his death in 1909. Where was he going in this intriguing picture, and where did he get the car?

75

1910 COUNT FRANZ GRAF HARRACH'S TOURER

A blood red car setting out for a drive in Sarajevo one fateful day in summer 1914. Among the feathered bigwigs were, in the back seat, the Archduke Franz Ferdinand of Austria-Hungary and his wife, the Archduchess Sophie. Within minutes of this photograph being taken, the royal couple were gunned to death by the Serbian revolutionary Gavrilo Princip. The First World War followed in the wake of this incident. This fateful car survives unscathed in the Museum of the History of the Army in Vienna.

76

77

76
1909 HOTCHKISS PROTECTED CAR
One of four "protected cars" ordered from France by the Turkish Sultan, and equipped with a rear-mounted machine-gun. Built to help the Sultan to quell riots, these early armoured cars were captured by the "Young Turks" and used against the Sultan himself. Note the crossed-cannon badge on the radiator. Hotchkiss had originally been founded by the American-born gun-maker, Benjamin Hotchkiss.

77
1916 MODEL-T FORD
Here's the IRA in flat caps ready to take on all comers. No attacking from the front, mind.

78
1936 FORD V8 PILOT
Take that, copper. A posed shoot-out on the mean streets of Depression-era England. The policeman has a bullet-proof window to shoot over. The villain is unwisely standing directly in his line of fire. Ideally, he should be racing away in his 90hp V8 Ford. Few of these sold in Britain before the Second World War; the Pilot made a strong comeback after hostilities ceased. They were popular, like Mk2 Jags in the 1960s, with police and thieves alike.

78

79

80

79
1958 VOLGA
A car much favoured by the secret police and security forces in the former Soviet Union and other Eastern bloc countries. They had a delightfully sinister appearance. Powered by strong 2.4-litre fours, they are slowly becoming collectors items. Volgas remain in production.

80
1965 ZIL
These enormous cars produced for top Communist party officials and, since 1991, for Russian leaders have traditionally been based on US models. The factory, founded in the 1930s, was originally called Zavod Imjeni Stalina, but the name was changed after the Soviet leader's death in 1953 to Zavod Imjeni Lihacheva. Lihacheva was the director of the plant. This Zil-3 would have been in its heyday during the early years following the overthrow of Nikita Krushchev by Leonid Brezhnev in 1964. The chrome-laden car was powered by an heroic, if not exactly proletarian, 200hp, six-litre V8. It could move.

81
1958 BMW 501
Leather-jacketed Munich policeman with mighty 3.2-litre V8 501 patrol car. These voluptuous cars were a mainstay of BMW car production from 1951 to 1964.

81

82

83

82
1936 LANCIA
The defeated Ethiopian emperor, Haile Selassie (1892–1975), being paraded through Addis Ababa by Mussolini's men in 1936. The Italians used gas dropped by aircraft to crush the Ethiopian army. The Italians were, in turn, deposed by a British army composed largely of Indian soldiers in 1941. Haile Selassie, who claimed to be a direct descendant of King Solomon and the Queen of Sheba, was reinstated, but deposed by Communists in 1974. He was murdered the following year. Rastafarians believe him to have been an incarnation of God. At least in this picture he is travelling in regal style.

83
1921 ALFA ROMEO TIPO RLS
Here is Benito Mussolini (1883–1945), a man constantly in a hurry, arriving at a holiday hotel at the wheel of his fast and furious Tipo RLS. The fascist leader was dictator of Italy from 1922 to 1943. His attempt to establish an Italian empire to rival the glories of ancient Rome and his allegiance with Adolf Hitler caused his downfall. Captured by Italian partisans in 1945, he was executed and his torso hung upside in public as a warning to would-be fascists and their sympathizers. It had been a very long road from that Italian hotel entrance and that glorious Alfa.

84
1936 VOLKSWAGEN BEETLE
The People's Car – initially the KdF ("strength through joy") wagen – was to have been sold to Nazi party members for 1000 Reichsmarks, or about £85. Some 80,000 hopefuls signed up for their Ferdinand-Porsche-designed, 100km/h, air-cooled, streamlined, autobahn cruisers, but Hitler decided to invade Poland, ignite the Second World War and, ultimately, bring destruction on his own people. They had to wait some while before they could save up again for new VW Beetles. Here, in happier days, is Reichmarshall Hermann Göring being shown the underside of a new VW cabriolet by some strong Aryan übermenschen.

85
1935 ZIS
Iosif Vissarionovich Dzhugashvili, 1879–1953, (aka Stalin, Man of Steel) ran the Soviet Union with a rod of iron more or less from Lenin's death in 1924 to his own, from a stroke, in 1953. In the meantime, he may have seen the Soviet Union win its heroic victory over Nazi Germany, but he also killed millions of his countrymen in savage purges and wilful starvations. Like all dictators, he was fond of his cars. Here he is inspecting the first Zis-101 limousines inside the Kremlin on April 29, 1936. He is flanked, on his right, by I A Lihachov, director of Zis, and G K Ordjonikidze, minister of heavy industry, and, on his left, by V M Molotov (foreign secretary) and A I Mikoyan (trade secretary). The Zis was powered by a 90hp 6766cc straight-eight. Its top speed was 115km/h.

84

85

86

87

88

86 + 87 + 88
1930s GROSSER MERCEDES
Adolf Hitler (1889–1945), *Time* magazine's Man of the Year, 1938, was devoted to Mercedes-Benz. He made the famous car company all but synonymous with his Nazi regime. He took a great interest in its cars, their design and engineering. Whenever, as in these three pictures, he was seen on public parade, it was in an open-top Mercedes. The cars were hugely powerful, all conquering, designed to last, all things that the Nazi regime was meant to be but thankfully wasn't. Designed to last 1,000 years, the Third Reich fell 988 years short of its target. The cars in these historic pictures have generally lasted much longer. The 770K Grosser Mercedes was prized by dictators and dodgy rulers worldwide. The Japanese emperor Hirohito ordered just the seven, while others were owned by General Franco of Spain, King Boris of Bulgaria and King Zog of Albania.

89
1936 MERCEDES-BENZ 540K
Adolf Hitler almost at the height of his power. This would continue to grow, like the output of the straight-eight motors of his Mercedes-Benz parade and touring cars, until Operation Barbarossa, the 1941 invasion of the Soviet Union. Hitler owned Mercedes-Benz cars from 1923, at the time of his failed coup in Munich, until his death.

89

90

91

90
1942 VW TYPE-166 SCHWIMMWAGEN
A parade of Porsche-designed four-wheel-drive amphibians based on the Kubelwagen, a military adaptation of the Volkswagen. The 166 was a popular troop carrier and capable of 80km/h, but was less popular in the water, where its speed was no more than 10km/h, and it was vulnerable to damage: a single .303 bullet could sink it. Some 15,000 Schwimmwagens were built.

91
1953 VW BEETLE
Lined up on parade outside Brunswick, these post-war Beetles still have the look of SS guards on parade at a Nazi rally. They were to take the automotive world by storm, and are still being made today in Mexico.

92

93

92
1961 LINCOLN CONTINENTAL
A lovely day now the sun has broken through in Dallas, Texas on November 22, 1963. Moments later, president John F Kennedy would be gunned down by Lee Harvey Oswald. Stylish President, stylish first lady, stylish car: sensational and sad day. Ever since, US Presidents have ridden in vast armoured motorcades, taking no chances.

93
1961 LINCOLN CONTINENTAL
President John F Kennedy slumps over in the back of the open-topped car after being hit by Oswald's bullet. One of the iconic images of the twentieth century.

94

94
MARTIN LUTHER KING'S
1963 RAMBLER AMERICAN
"I have a dream that my four children will one day live in a nation where they will not be judged by the colour of their skin but by the content of their character." This is Dr Martin Luther King (1929–68), Baptist pastor and civil rights champion with his son Martin Luther III, aged seven, walking up to their house in Georgia, Atlanta after church. Dr King drove a modest car but also drove an ambitious campaign for human rights in the United States. He was shot dead on the steps of the Lorraine Hotel, Memphis, Tennessee on April 4, 1968 by James Earl Ray.

95

96

95
1953 DAIMLER EMPRESS
The royal family is said to have stopped using Daimlers in 1950, but here is the young Queen Elizabeth, with Prince Charles and Princess Anne in tow, at the wheel of what is surely a Hooper-bodied Daimler Empress, probably on a DE24 chassis. Whatever the truth, the Queen looks happy to be driving it. The Queen is a blood relative of the former kaisers at the time of the founding of the Daimler car company in Germany; the British Daimler company is related only by association to what since became the Daimler-Benz empire. Frederick Richard Simms, an Englishman, met Gottlieb Daimler in 1880 and, later, created a subsidiary of sorts back home. In practice, the two companies were entirely separate.

96
1958 LAND ROVER SERIES 1
HM Queen Elizabeth and HRH the Duke of Edinburgh on board HMS *Albion* in a Land Rover in 1959. The Land Rover, based on the wartime American Willys Jeep, was only meant to be an Austerity-era stop-gap model to be sold in the colonies. Instead, it became a national favourite, beloved of farmers and the Queen. HMS *Albion* was launched in 1947 and completed in 1954. Weighing 20,000 tons unladen, and powered by 78,000hp Parson's steam turbines, she could cruise at 20 knots for 6,000 miles non-stop. She saw action in Suez, Aden and Indonesia, was converted to a helicopter carrier at the time of the Queen's visit and was broken up in 1973. Land Rovers, including many Series 1s, are still very much with us. The Queen and the Duke of Edinburgh, too.

97

98

99

97

1954 LAND ROVER SERIES 1

Here is Winston Churchill (1874–1965) with his trademark Cuban cigar and his favourite Land Rover. The legendary wartime leader was Prime Minister for a second time (1951–55) when he took delivery of his Land Rover. The company had sold a record number of cars in 1954, but Churchill's support was to boost sales further. The Land Rover certainly displayed the "bulldog" characteristics that had so endeared Churchill to the nation as a wartime leader.

98

1943 WILLYS JEEP

The legendary General Purpose – GP, and so "Jeep" – from Willys, one of the all-time great cars. It did much to help British, American and Commonwealth troops during the European campaigns of the Second World War. Here Field Marshall Montgomery, commander of the Eighth Army that defeated Field Marshall Rommel in the north African desert, is chauffeured by his regular driver J Burford. They don't appear to be having much fun.

99

1943 WILLYS JEEP

President Franklin Delano Roosevelt greets US troops in Morocco in January 1943. Everybody got to ride in a Jeep, from a private to the President.

100

101

102

100
1967 FORD MUTT
"I love the smell of napalm in the morning"… US troops attempt to "completely level the Vietcong stronghold of Ben Suc" by torching the place. More than two million Vietnamese civilians died during the Vietnam War, along with some 58,000 US soldiers.

101
2003 LAND ROVER
Soldiers of the heavy machine gun platoon of the 1st battalion Irish Battle Group checking out a burning oil well during the US-UK invasion of Iraq, March 2003.

102
1991 AM GENERAL HUMMER
These vast off-road vehicles, seen here lining up on patrol in Saudi Arabia before heading to Iraq, were designed, from 1979, for the US army. Prototypes of the HMMWV (High Mobility Multipurpose Wheeled Vehicle), or Humvee or Hummer, were tested in 1982. The giant vehicle went into volume production in 1985. A civilian model followed after the publicity gained in the Gulf War in 1992. It became a cool "gangsta rappa" style accessory. Some accessory, and, to be honest, not much fun to drive even on the meanest city street. It comes into own well off the road, and, mostly, on the battlefield.

3253

MINI CARS

In 1945, much of Europe was in a sorry state. In Britain, which had lost an Empire but won the war against Hitler, Mussolini and Hirohito, with more than a little help from its American and Russian friends, ration books endured until summer 1954. At the nadir of the national austerity programme pursued by Clement Attlee's Labour governments of 1945–51, most car owners were allowed just five gallons of petrol per month. This would carry a toff in a Rolls-Royce Phantom III, mothballed during the war and now, creakily back on the road, some 40 miles. But Lawrie Bond (1907–74), a miniature car specialist from Preston, could offer austerity drivers 1,800 miles behind the wheel of his MkA Minicar. Not only were cars rated at under 9hp eligible for a happy 18 gallons of petrol, but Bond's first three-wheeler, using lightweight construction methods drawn from the aircraft industry, also promised 100mpg.

The first production car was driven down the 228 miles from Bond's Lancashire factory to London in March 1949. The 122cc, 5bhp Villiers motorbike engine saw the little aluminium open-topper cruising at 30mph. The average speed for the journey was 22.8mph and the fuel consumption worked out at 97mpg. By the time Bond's much improved four-seater "family model" 197cc MkC model went on sale in 1952, it faced competition from the 346cc AC Petite and 747cc Reliant Regal. The Bond was still the cheapest of the austerity mini cars, but the competition, if not exactly quick, was hotting up.

Stirling Moss, the already famous racing driver, nodded his approval for the Bond Minicar in a magazine advert. Despite heavy snow and freezing fog, two regular British army officers, Lieutenant Colonel Crosby and Captain Mills successfully completed the 1954 Monte Carlo Rally in their MkC. The journey, from Glasgow, took three and a half days, at 63mpg. In 1959, Douglas Ferreira, a Bond employee, drove a 250cc MkF from Land's End to John O'Groats in 23 hours 40 minutes at averages of 36.89mph and 60mpg.

I got to drive a MkF when I was 17. I thought I wanted one. It was a funny little thing, surprisingly nippy, capable of 55mph, and this one boasted a hard-top and reverse gear (able to turn in their own length, Bonds never really needed reverse). But my young motoring heart was stolen by the Mini. The Mini was nowhere near young then, but, compared to the Bond, there was simply no competition. Where Alec Issigonis's Mini remains a milestone in motoring history and a true design classic, the Bond Minicar went the way of the AC Petite, Reliant Regal and pretty much every other abstemious miniature car designed to get around the shortage of petrol in 1950s Europe.

The variety, ingenuity and sheer chutzpah of these funny little cars continues to make us smile. Throughout Europe, manufacturers – particularly those with expertise in aircraft and motorcycle manufacture, designed and made their own takes on the Bond. Isa's Isetta was the first of the famous Italian bubble cars, later made and developed under licence by BMW (36,000 built between 1955 and 1962). Heinkel and Messerschmitt, two names that had struck fear into the souls of millions in Europe as their potent warbirds attacked targets from the North Downs of Kent to the Russian Urals, manufactured their own bubble cars based on the perspex canopies of military craft, although without the 20mm Hispano canons that had once been mounted in front of them.

Willi Messerschmitt and Fritz Frend's joyous little KR175 "Kabinenroller" was, stylistically, his Bf110 fighter-bomber canopy in the guise of a car, complete with tandem seating and joystick steering column. Its 9bhp, 174cc Sachs two-stroke single-cylinder motor was slightly down on power compared with the aircraft's twin 1,100hp Junkers Jumo engines, but in 1957 Frend launched the four-wheeler, 494cc twin-cylinder TG500 model. The "Tiger" is, as I have discovered, a revelation. It is fast (130km/h maximum speed; not that I tried this) and handles extraordinarily well. There is no comparison between the engineering sophistication of this truly sensational car and contemporary British mini cars. On the classic car markets, Tigers are highly prized.

Marvel though it was, the Tiger was expensive. In 1959, the Mini arrived and, suddenly, all these post-war motoring curiosities seemed completely outdated. Only one other mini car came anywhere close to the Mini in terms of cult status and sales. This was the lovely little Fiat 500 designed by Dante Giacosa and launched in 1957. The first car I ever drove was one of these Torinese gems. It featured a hand throttle. You could roll back the sunroof, sit on top of the driver's seat, set the throttle and steer with your feet. In bare feet and on a beach only, of course, officer.

Even though the Mini dominated the miniature car market for most of the 1960s, rivals continued to be developed. Honda's sweet-running N360 and N600 saloons led on to one of my all-time favourite sports cars, the tiny 1966 Honda S800. Its high-revving engine is exquisitely engineered. Displacing just 791cc, it boasts twin overhead camshafts rotating on needle-roller bearings. This allows it to spin, like a motorbike, safely up to 8,500rpm. With 70bhp on tap, refined suspension and sharp steering, this jewel-like example of bonsai engineering is all but priceless. S800s were both highly prized and raced successfully in the United States. They made otherwise handsome, characterful and entertaining contemporary MG Midgets and Austin-Healey Sprites seem like machines from a quite different age.

Today, the mini car market is booming. Although the BMC Mini itself has grown into a surprisingly big BMW, the Smart Car, engineered by Mercedes-Benz for a Daimler-Chrysler subsidiary and launched in 1998, is very small indeed. It has, though, great presence on the road. With a turbocharged three-cylinder 599cc Mercedes engine, it goes well, the cooking model limited to a top speed of 84mph. The Smart can be parked nose to the kerb like a motorbike. Soft panels means that it can cope with the bumps and scrapes of city motoring. It is a measure of how Europe has changed; our city streets might still be narrow, our city centres densely populated, but, although now affluent, we have learned to make mini cars again, this time as much fun to drive as they have always been to look at.

103
1963 PEEL 50
The Peel was so light that it could be picked up and turned around by any young London Miss; in fact, this was how the car was reversed. Enough said. Fewer than 100 of these tiny British single-seaters were built. They were powered, if that's the right word, by a single-cylinder 49cc Zweirad Union (DKW) two-stroke. Performance was, inevitably, nugatory.

104
1955 TROJAN 200
After the Second World War, German aircraft company Heinkel, better known for its Do-217 "Flying Pencil" bombers, made these three- (and four-) wheeled bubble cars at Spayer, as well as in factories in Ireland, Argentina and England. Among the more civilized micro cars, they were nicely engineered and had 10hp 198cc engines. Now whether or not this cautious Fräulein is saying hello boys! or you must be joking, chum, it is hard to say.

103

104

45
AMBULANCE
HEG 175

106

107

105
1955 BMW ISETTA 300
Cyclist about to overtake a Ford Transit-based ambulance held up on an English country lane by a delightful example of *das rollende ei*, BMW's "rolling egg" bubble car. Not much fun at over 30mph, these German-built Isettas – the original cars were designed and built by Isa, the Italian motorcycle manufacturer – can beat 50mph, but you would have to be brave to aim for such giddy heights. Built under licence from 1955 to 1964, 160,000 examples of the 300 were made in Germany. Nearly all are four-wheelers, although the close-set rear pair means that they are routinely mistaken for three-wheelers. Engines are either a 247cc or, more normally, 13hp 298cc single-cylinder BMW motorcycle unit. Like all micro cars, sales were hit hard by the arrival of the BMC Mini in 1959.

106
1955 ISETTA MOTOCOUPE
How to travel in bubble car in style. Employ a chauffeur. Dress well. Wave from the sunroof while pretending to be Princess Margaret having a laugh. Do not exceed 30mph if you want your hair-do, earrings or sanity to stay in place, or one piece.

107
1955 BMW ISETTA
Let's do lunch! Smart lunching ladies get aboard a 247cc Isetta at the car's launch at the 1955 Earl's Court Motor Show.

105

108

109

110

108
1957 DKW JUNIOR
DKW went on to become Audi and the rest is a very successful history. This little American-style two-door coupe is the prototype DKW Junior, a twin-cylinder 660cc two-stroke mini car. Well engineered, although inevitably a little smoky, it went on to sell 118,968 examples.

109
1959 OPPERMAN STIRLING FAMILY SPEED SALOON
Designed by Lawrie Bond of Bond Minicars, the Stirling was a rather smart-looking two-door glass-fibre coupe that looked much like NSU's little Bertone-designed Sport Prinz. It could seat two adults and two children and, with a 424cc Excelsior two-stroke twin, could scurry up to a claimed 70mph. But it cost £541 7s, which meant it faced direct competition from the new Mini. Only two were built; one survives.

110
1959 NSU PRINZ 30
Who does he think he is? Stirling Moss… driving a German mini car. Moss, the great British racing driver, has always been fond of micro cars and motor scooters, the automotive opposite of the furiously fast racing cars he has piloted for more than 50 years. Here he is with a 1959 NSU Prinz sporting his own, personalized numberplate. The car had a 598cc 20hp twin. It would win no races.

111
1936 FIAT 500
No joke at all. This is a British-registered model with a sunroof in the rain at a Welsh rally in 1937. The 500, or "Topolino" ("Little Mouse", after Walt Disney's Mickey Mouse), was a superb little car, engineered by the motoring giant Fiat to standards that most of the independent post-war mini car makers were quite unable to match. The car was engineered by the brilliant Dante Giacosa during 1933–34 (he was just 28 years old when commissioned to undertake the project) and styled in-house by Fiat's Rudolfo Schaffer. The result is a very fine car. The engine is a 569cc in-line ohc watercooled unit with two valves per cylinder. The car features hydraulic brakes, 12-volt electrics, independent front suspension and a top speed of 85km/h. It handles well and rides like a much bigger car.

111

112

112
1957 FIAT 500
Twenty-one years later (see previous page), and the 50-year-old Giacosa produced a second mini triumph, the much-loved "Nouva 500" or"Cinquecento". It looked good in every setting, from the ducal square here in Turin to cobbled streets in every hill town you can think of. Approximately 3,678,000 were built in a number of guises, including an estate version, a beach "buggy" and a Topolino-lookalike that was also just like the illustrations for Noddy's car in Enid Blyton's children's stories. This time the car's engine was air-cooled – an enthusiastic 479cc twin – and mounted behind the tiny, four-seat cabin. The car could top 85km/h happily, and all day. Thousands roam the streets of Italian cities today.

113
1972 CLOCKWORK ORANGE
Pipped at the post… the Outspan orange advertising car squeezes to a halt at a zebra crossing in London. This funny car was based on the floorpan and running gear of a BMC Mini. Its job was to advertise South African oranges at a time when South Africa's policy of apartheid had made it one of the world's least liked regimes. The car survives in the National Motor Museum at Beaulieu, Hampshire.

Snuff Box Doubles Bar
Shorts OF LONDON
THE LANES CELLAR BARS
The Amazing
OUTSPAN
Orang

AIL CRISIS LATEST
FIRE
COUNTY
BARCLAYS BANK LIMITED

114

115

114
1955 MESSERSCHMITT KR200
Like an Me-109 cockpit in search of wings, the glorious KR was a fine, fast two-seater. Here one buzzes through Piccadilly Circus trying to outdrag a quad of handsome FX3 taxis, a Standard Vanguard and a Morris Oxford. The buses appear to be on strike, so perhaps this is a scene from 1958 when the red buses disappeared for weeks on end. The KR200 was capable of 100km/h courtesy of a 191cc Fichter and Sachs 10hp single-cylinder, two-stroke engine.

115
1952 FEND FLITZER
Carefree and not out to impress anyone, a young chap whizzes along in a single-seat, KR200 prototype.

116

116
1954 TYPE-C BOND MINICAR
Oh brave new world that hath such people in it… Ration books have just ended, so this happy family have decided to celebrate with a Bond Minicar. Stuff safety belts and all that palaver, they're motoring in true micro car style.

117
1960 NOBEL 200
Foxy lady seeks Noble. Noble, not Nobel. Aagh!

118
1967 HONDA N360
A thoroughly engineered Japanese rival for the well-established British Mini, the front-wheel-drive Honda N360 was neatly designed and drove well, with power – 31bhp, just 3bhp less than an 850 Mini – from a free-revving and jewel-like 354cc ohc twin. A larger 598cc was available from 1968, encouraging some export sales. But this was little more than an Oriental curiosity in the West at the time.

117

118

119
1983 SUZUKI CV
A 49cc retro-style mini show car by the motorcycle giant.

120
1967 SUZUKI SHOW CAR
Things must have been getting desperate. Mine's a Mini.

121
1957 GLAS T700
Hans Glas made his name with his popular Goggomobil micro car, but for those on their way up to something a little bigger and more like a real car, what about this charming, two-tone American-style sedan of 1957? The Fräulein in polka dots is listening out for the 688cc engine. Is it at the front or the back? That would be telling, says Fritz. Now kindly get off my bonnet; I can't see where I'm going. This pretty car was in production until 1965.

119

120

121

122
1967 MINI
How many 1960s fashion victims can you fit in a Mini? Who knows, man, I wasn't there. Fourteen is the correct answer, in 1966, if you must know.

123
1966 MINI COOPER S
Cooper S Minis won the tough, long-distance Monte Carlo in 1964 and again in 1965. When Timo Makinen won again – seen here battling through the snow in "GRX 555D" – the French authorities declared the win null and void. The Mini had broken the rules. How? It had the wrong sort of headlamp bulbs, said the absurd, and xenophobic, officials. It was sweet revenge when a Cooper S won again in 1967. The Cooper S was a formidable car.

124
2002 MINI COOPER S
The new Mini is a chunky car designed for the brawny streets of US cities, such as Times Square in the driving rain. It has been a big success in New York. Except in terms of its retro-aesthetic and general layout, it is a very different car from its sparky, if often poorly built, predecessor. Although built in England, the new Mini has been engineered with considerable BMW input and know-how.

122

123

124

125

125

1965 MORRIS MINI MINOR

The Mini was truly a classless and go-anywhere car, although blue-collar workers tended to prefer bigger, more conventional and more American-style Fords. To cover all class and age profiles, early Minis were badged Morris, Austin, Wolseley and Riley. The last two were meant to appeal to older, more conservative generations. Winsome and all but twee, they hold great appeal to Japanese collectors today, but very few are seen on British roads; many not so common, nor so garden, basic Minis are.

995 RKO
ACE 115B

CARS AND ARCHITECTURE

The Chrysler Building. Spaghetti Junction. The Fiat factory in Turin. Executive housing estates with triple-garage houses. Your local shopping mall. The car has unquestionably reshaped architecture, civil engineering projects and our urban and rural landscapes. Many famous Modern Movement houses by the likes of Le Corbusier (1887–1965) and Frank Lloyd Wright (1867–1959) were designed around car ports and the turning circles of cars. Wright, America's most famous architect, was obsessed with cars. His favourite was probably his coffin-nosed, Lyoming V8-powered '36 red Cord 812 convertible, Gordon Buehrig's design masterpiece. As for Le Corbusier, Europe's most celebrated architect of the past century, his Voison featured in hundreds of photographs of his radical buildings. The architect's famous Modernist polemic, *Vers une Architecture* (translated as *Towards a New Architecture*, 1923) was crowded with illustrations of machinery revealing its essential and "honest" beauty; the latest cars took pride of place alongside new-fangled aircraft.

In fact, it can be argued that the low, horizontal streamlined form of many owe their lines to the car. In eighteenth century cities, the rhythm of rows of tall, thin Georgian houses took its cue from the pace of someone walking past them. Long, scenographic Regency terraces, like those surrounding Regent's Park in London, took their cue from the pace of a horse and carriage. The extended horizontals of buildings like the Art Deco Hoover factory, designed by the American architect, Wallis Gilbert, on London's Western Avenue, or long, low prairie houses by Frank Lloyd Wright take their cue from the speed of the car. As cars went faster, so modern buildings appeared to abandon the kind of decoration and detailing that only someone walking past would really notice. If you flashed past in your Cord or Voisin, such leisurely details would be both invisible and all but meaningless. I am not sure if this is entirely true, yet it is something I have often felt as I have driven past motor-age buildings.

What we know for sure is that the car has affected the plans and structure of many buildings. Car parks built under or even up into the superstructure of twentieth century buildings had a radical impact on the way they looked and how they were used. In downtown Chicago, there is a wonderful pair of 1960s apartment blocks – Marina City – that, famously, resemble giant corn cobs. Designed by Bertrand Goldberg (1913–97) each comprises 450 apartments and 450 parking spaces. The latter occupy the bottom third of each block; cars reach them up a continuous concrete spiral. This not only keeps the cars off the busy streets below, but it also means that their owners are not forced into underground car parks, which, no matter how well designed – as, for example, they are in the centre of Lyons – are always a little disturbing and unpleasant. As a bonus the towers are wonderfully animated by night as cars, headlamps shining, climb up and down the circling ramps. Here, the cars enhance rather than detract from the buildings.

The most famous example of this, though, must surely be the Fiat factory in Turin. Built from 1916, by a team of architects and engineers led by Giacomo Matte Trucco, the Lingotto factory was based on design principles adopted from Henry Ford's factories across the Atlantic. What made the Fiat building so very special was the roof-top test track. Looking at photographs of cars whizzing around this great roof-top circuit with the great baroque city beyond, it is easy to see Trucco's masterpiece as an embodiment of the Futurist ideals drawn up by the poet Marinetti in 1909. Here, truly, was the concrete embodiment of a modern, technological dream: cars, architecture and the city working in dynamic harmony. Today, the factory has been converted into a number of museum, galleries, restaurants and public performance spaces for the city of Turin by the Genoese architect Renzo Piano, who, incidentally, once designed an experimental car for Fiat in the mid-1980s.

Car parks are the most obvious car-determined form of building; these vary, worldwide, from the gross and offensive to some with real charm and some that are even much loved, if not exactly for their design. Among these is the multi-storey 1960s car park in Gateshead, in north-east England. A campaign to save this brutalist concrete structure got under way in 2002 when it was threatened with demolition. Why? Because it starred as a backdrop for a dramatic scene in the cult British gangster movie *Get Carter* (Mike Hodges, 1971). Today, the car park is a shrine for film buffs.

The car has encouraged spectacularly lazy forms of places and buildings, notably the drive-in movie theatre and the drive-thru' burger joint. It has even moved into people's homes, such is our love of the car and our need to be near it at all times. A house built in 1998 by the London architect Seth Stein for a client in Knightsbridge incorporates a smart glass lift that raises its owner's cherished canary yellow Fiat 500 into a gallery-like space on the first floor, where it can be admired while guests eat dinner. But as the best looking cars are "rolling sculpture", they deserve gallery space as least as much as many conventional artworks. When Buddy Rich, the legendary jazz drummer, was asked on the BBC's "Desert Island Discs" radio show which one luxury, along with his eight favourite gramaphone records, he would take to his desert island, he chose his beloved Ferrari Daytona. Told by the programme's presenter, Roy Plomley, that there would be no petrol on the island, Rich said he would be happy just to look at the shape of the shark-like Bertone-designed beauty.

Where cars and architecture, cars and cities go wrong is when they take over the place and turn every last plaza, piazza, square and circus into makeshift car-parks. Some argue that cars add a sense of life and vibrancy to city centres, which up to a point, of course, is true; yet who could possibly step into the great courtyard of the newly restored eighteenth century Somerset House, by Waterloo Bridge in central London, and look back wistfully to the days when this magnificent public space was filled with government employees' cars? Cars, however beautiful, do need to know their place.

126

127

126
FORD MOTOR COMPANY MODEL-T FACTORY, DETROIT, MICHIGAN
Today, most of the early Detroit automobile factories are abandoned ruins. This is Ford's Model-T factory at the height of the car's fame and sales success. The photograph captures some, at least, of the 50,000 employees who laboured on the assembly lines founded here in 1913. Henry Ford's greatest architectural asset was Albert Kahn. Born in Westphalia, Germany, Kahn was to build more than 600 factories, 521 of them for Stalin in the Soviet Union. The architect of the assembly line, he created daunting, if seemingly effortless, buildings stripped almost of all decoration. These captured the imagination of writers, film directors and modern European architects including Le Corbusier. They were powerful and unforgettable symbols of sheer industrial might. The car, and, above all, Ford, played a key role in their development worldwide.

127
FORD ROTUNDA, DEARBORN, MICHIGAN
Albert Speer, eat your heart out. Designed by Albert Kahn for the 1933 Chicago World's Fair, this dramatic exhibition pavilion was deconstructed and shipped to Ford's River Rouge Dearborn factory in 1936. It was used to show new cars, as here, for corporate events and parties and for Ford's children's Christmas season each year until it burned down in 1962. At one time the Rotunda was the fifth most popular tourist attraction in the United States.

128
FIAT LINGOTTO
FACTORY, TURIN
Speeding into space. Fiat test drivers race around the great roof of the magnificent five-storey concrete factory that Giacomo Matte-Trucco designed for Fiat in the early 1920s. This scene dates from 1929. The factory was self-consciously modelled on Albert Kahn's work for Ford in Detroit. Le Corbusier was deeply impressed by this machine age masterpiece; he went so far as to describe it as "a guideline for town planning". Imagine anyone saying this in the age of the Congestion Charge. The factory no longer makes cars; it has been converted by the Genoese architect Renzo Piano, who once designed a prototype car for Fiat in the mid-1980s, into performance, exhibition and meeting spaces. The Lingotto factory retains its place in the heart of industrial Turin.

128

FIAT

129
AMERICAN SUBURBIA, 1952
In giving us freedom, the car also gave us the ever expanding city. So much so that, as early as the mid-twentieth century, suburbia has blossomed across – or blighted, depending upon your point of view – vast tracts of Britain and the United States. Perhaps this mattered less in the US, where land, petrol and cars were all cheap compared to little, precious and costly Britain. With the car came the garage, wider streets and the decline of public transport. In terms of urban planning and architectural design, the car gives and the car takes away.

130
ROUTE 4, ENGLEWOOD, NEW JERSEY, 1935
As Fritz Todt engineered his magnificent autobahns in Germany, so the US federal government pioneered its seemingly boundless freeways. The car scythed its way unhindered through landscapes worn and virgin. Here in New Jersey, attempts were made to civilize the new roads with smartly uniform street lamps, generous tree-planting and pedestrian walkways. The scene seems less innocent today, cars grinding along nose to tail. Back in the 1930s, these roads seemed heroic; nothing quite like them had been built since the Romans.

129

130

131

132

131
SPAGHETTI JUNCTION, BIRMINGHAM, ENGLAND
There is a website called "How to Avoid Spaghetti Junction". This complex weave of concrete carriageways was never meant to be a nightmare. Designed by Owen Williams, engineer of Britain's first inter-city motorway, the M1, the US-style junction was meant to free traffic and not to twist it into a grid-locked tangle. There are times when this 1960s megastructure is quiet, but it lost its "white heat of technology" innocence years ago. Note how the heroic structure of the road junction contrasts with the dinky suburban housing estate alongside – a very English scene.

132
SEATTLE, 2000
Here, at the start of the twenty-first century, is the American city and its intimate relationship with the road and thus the car. Few developed countries are so dependent or quite so in love with the car. The scale of such roads defies common sense. Cars are fine in their place, but let's have some railways for heaven's sake. The US used to have some of the world's best long-distance trains. These were killed off by the end of the 1950s by cars rumbling along optimistic freeways.

133

133
DRIVE-IN MOVIE THEATRE, COPENHAGEN, 1961
There are something like 550 cars parked here on the opening night of Scandinavia's first drive-in movie theatre. How many can you identify? The theatre might look like a quarry to you and me, but this was glamorous stuff in 1961, adopted from the US.

134
DRIVE-IN MOVIE THEATRE, LOS ANGELES, c 1945
This is the original, or one of them: an LA drive-in when Bogart and Bergman were among the stars of this very big screen. Note the loudspeakers blaring into the cars. The drive-thru burger restaurant, bank and whatever else were to follow soon enough.

134

135
US CAR PARK, c 1940
A sea of veteran Americana. It was a delight for car enthusiasts, but a nightmare for many people as the car began to eat up land hungrily from the 1930s. By the way, who says that cars all look the same today and didn't then?

135

B LANE

IN
OUT
UFJ
980
777 JHY

136

MULTI-STOREY CAR PARK, BRISTOL, 1960

Parking spaces in British cities were always going to be a problem; as the number of cars sold rocketed during the consumer boom of the late 1950s, so local authorities began investing in multi-storey car parks such as this 550-space concrete behemoth. Although often despised, these were necessary structures if the car was to be king, queen and all princes in our city centres. The delightful thing here is the contrast between the spindly looking cars – a Ford Anglia and a Hillman Minx estate – and the Space Age heroics of the architect-designed garage they creep in and out of. The men in white coats have long gone.

137

AUTOMATIC CAR PARK, SOUTHWARK, LONDON, c 1961

This was one way of cramming 464 parked cars into a restricted space: raise and slide them into position on steel racks. This one has just opened near London Bridge, shortly before it was sold off to an American buyer and replaced by a new one. The period cars include a Riley 1.5, Austin A40, Hillman Minx, Ford Consul Mk2, Jaguar Mk2 and Vauxhall Wyvern.

137

138

138
1955 CITROEN DS19
Much like the exquisite sixteenth century French tapestry depicting the Unicorn in Captivity, on show in the Cloisters museum, New York, Flaminio Bertoni's superb Citroën DS sits captive in a circular compound fenced off from the prosaic cars around it. This was the Paris motor show, 1955. By the end of the first day, Citroën had taken 12,000 orders for the car, proof that the public is not always as old-fashioned as it is often made out to be. Bertoni (1903–64) had designed the Citroën Traction Avant as well as its 2CV. The DS – *deese*, or goddess – was his third masterpiece. Its qualities are as much architectural as they are mechanical; it inspired a new generation of architects, designers and even philosophers. "I think that cars today," wrote Roland Barthes, "are almost the exact equivalent of the great Gothic cathedrals: I mean the supreme creation of an era, conceived with passion by unknown artists, and consumed in image if not in usage by a whole population which appropriates them as a purely magical object."

139
SHOPPING MALL, SAINT LAURENT, FRANCE, 1970
No car, no shopping mall. The two have developed hand in hand. Here, in a wonderfully bland new French mall, the cars have come inside. Why not? They belong here as much as "Byron", "Singer" and "Timwear". The cars on show are the new, air-cooled Citroën GS mixing it with the last generation of the epoch-making Citroën DS and the bouncy little Ami.

galeries
byron
tricothèque
Timwear

140

141

142

140
MONT BLANC TUNNEL, FRENCH-ITALIAN ALPS, 1965
Not even the Alps were a challenge for the car. Hannibal had crossed them with elephants in 218BC, but, from July 16, 1965 families in sturdy Fiat saloons like this one could drive straight through them. At 11.6km, the tunnel was the longest of its kind in the world. It was closed for three years after a devastating fire caused death and much damage on March 24, 1999.

141
OAKLAND BAY BRIDGE, SAN FRANCISCO, 1936
Road bridges are an art never quite perfected, but always stirring and often magnificent. This great steel suspension bridge over Oakland Bay was opened by President Roosevelt on November 12, 1936. The President has only just cut the ribbon here as a stream of cars rumbles over the brave new $77.6 million structure. It was considered one of the new wonders of the world at the time.

142
UNDERPASS, MANHATTAN c 1960
By the 1960s, cars were passing over, under, below, between and beside the streets of our cities. Like ants on the march, nothing, it seemed, could stop them. Here cars plunge down from New York's city district. City Hall is the backdrop along with all-American billboards for Coca-Cola and Pall Mall cigarettes, long before the Surgeon-General had determined that they were bad for your health. Before seat belts and emission tests, too.

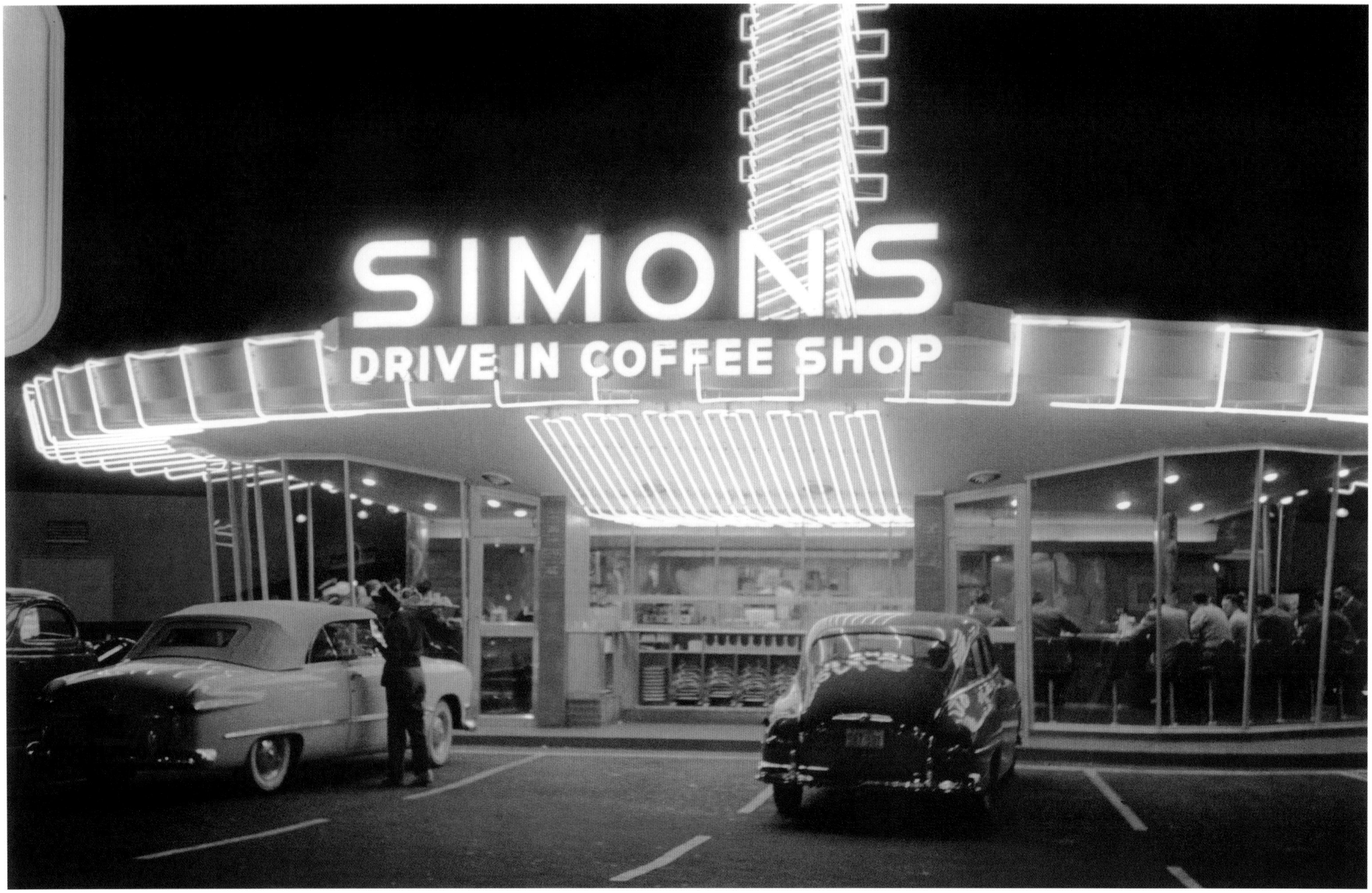

143

143
SIMON'S DRIVE-IN COFFEE SHOP, LOS ANGELES, 1951
This drive-in was in Hollywood. It took its design cue from neon-lit movie theatres. You could either sit inside, like models from an Edward Hopper painting, or be served by a smartly uniformed waitress without moving your ever bigger butt from behind the wheel.

144
GAS STATION, WENDOVER, UTAH, c 1945
This shot captures the notional romance of long-distance driving, US-style. The cinematic gas station offers not just fuel for the car, but fuel for the inner man, too, and "cabins" to rest up in for the night. It has its own special, low-rent glamour.

145
GAS STATION, MINNEAPOLIS, MINNESOTA, 1937
The speed of the car encouraged forms of architecture that could be seen and identified from a distance, and by night. This spectacular 1930s gas station was built almost entirely in glass bricks so that it shone like a single giant lamp at night. A delightful conceit.

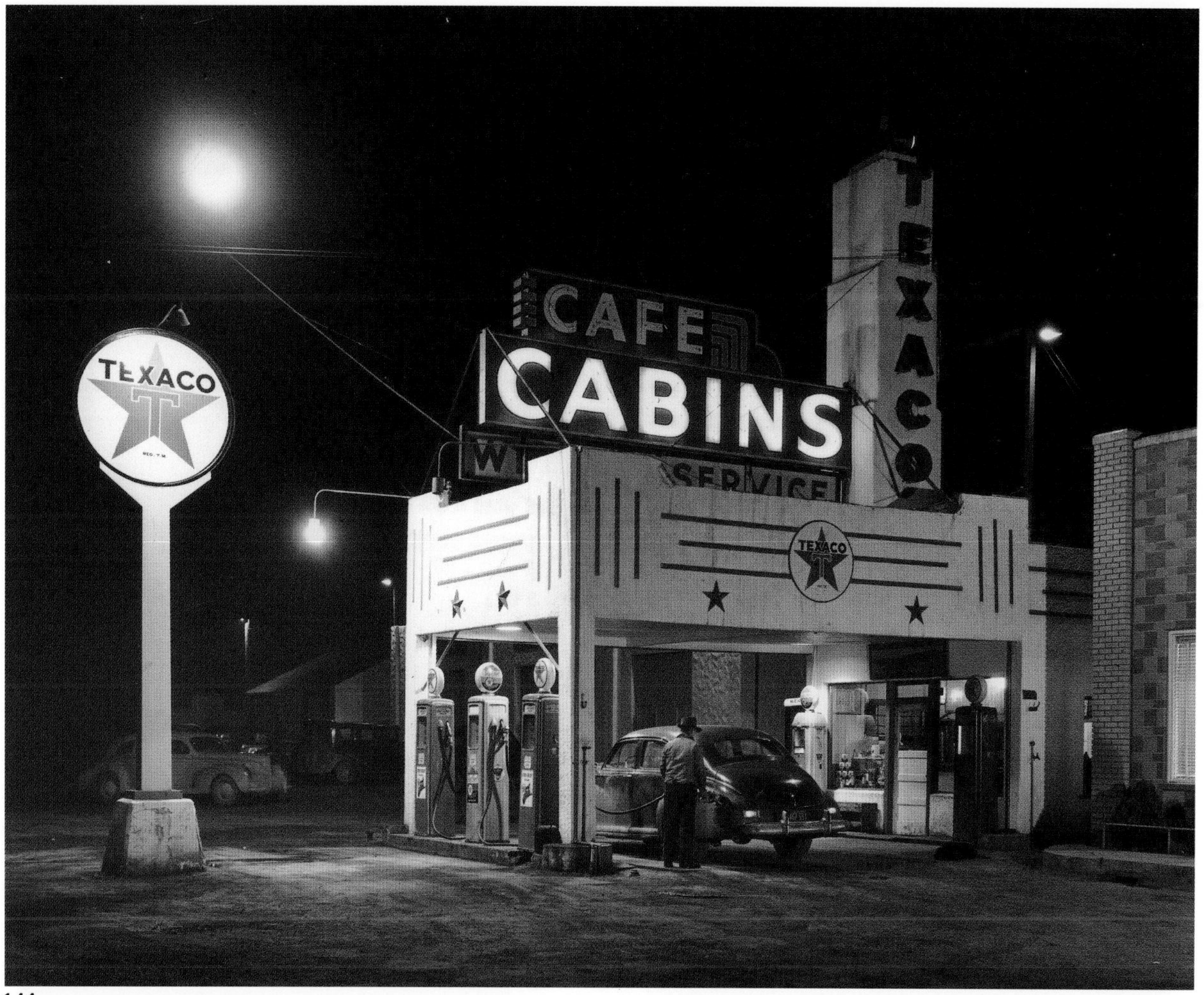

144

145

146
PETROL STATION, BLASHFORD, HAMPSHIRE, 1930
That'll be ye olde petrol station, sir. Odd of you to be driving a foreign car, sir. What is it? A Renault, you say… A venerable thatched petrol station complete with leaded-light windows dating, it seems, from the sixteenth century. Or, how to motor the English way even when driving a car built by Johnny French. Long gone, sadly.

147
GULF LIGHTHOUSE SERVICE STATION, MIAMI BEACH, FLORIDA, 1937
Lovely period artwork for this new, nautical-style service station inspired clearly by the speed and shape of cars, as well as its setting by the ocean. It offers not just gas, air and water for the car, but also a hotel with restaurant and cocktail bar. Saves drinking and driving then; not that many drivers seemed to think this was a problem in the 1930s.

148
PONTIAC SERVICE STATION, MEXICO CITY, 1942
Great combination of Mexican hacienda-style architecture and unconstrained neon advertising hoarding. The car was very much king at this time; no apologies need to be made for it, neither in the amount of fuel it used nor in the way buildings designed to serve it affected the look and feel of city streets.

146

147

Pontiac
CORNEJO
AUTORIZADO por GENERAL MOTORS
FERIA DEL JUGUETE
Pemex-Penn
PEMEX
EQUIPOS MODERNOS S.A.

148

KELLY
DUNLOP
WELLWORTHY
SMITHS STAMPINGS
Peugeot
SINGER
MG
JAGUAR
MORRIS
TRIUMPH
AUSTIN
HILLMAN
WOLSELEY
STUDEBAKER
ROOTES GROUP
FORD

CARS AND ADVERTISING

One of my favourite advertisements here is for Bugatti (page 164). It shows a very stylish young lady about to drive off in her black and yellow Type 57, one of the finest cars ever built. There are no words. Correction. There is just the one. In the bottom right-hand corner is the legend "Bugatti", a badge of unqualified excellence. It seems a wonder that Bugatti bothered to advertise at all. Word of mouth, or simply the sight of those matchless cars, or the sound of their exhausts – tearing calico; it's always "tearing calico" with a Bugatti exhaust – would be enough, you would have thought, to have got the cheque books out in the 1920s and 1930s.

The point here is that Bugatti needed no fine words to butter mechanical parsnips. Its cars were meat and vegetables, pudding, fine wines, champagne and brandy, too. Lesser marques have needed to sell their cars hard, and often, although not always, there is a rule of thumb that works: the worse the car, the fancier the advert needed to sell it.

The unintended humour of the advert for the Ford Edsel on page 167 is delightful. "They'll know you've arrived when you drive up in an Edsel." They certainly will. They will also run a mile or crack up laughing. The poor old Edsel – not such a bad car, really – was one of the biggest flops in motoring history. Subject of an unprecedented $250 million advertising campaign, the car was launched just as the US economy was taking a nose-dive. No one, not even most of Ford's management, liked the name (it belonged to Henry Ford's son) and many people actively disliked the car's trademark "horse-collar" grille. Whatever the reasons not to buy, Ford sold just 100,847 Edsels – again, not that bad – during its production run between 1957 and 1960, with a further 7,431 sold in Canada. It had banked on selling 200,000 in 1957 alone.

The advertising slogan for the Mk C Bond Minicar is equally, if less disastrously, funny. "You too ought to own a Bond Minicar," it says. Note the "ought". Perhaps you *ought* to, but one look at the curious cartoon-style car says you just might not want to own one.

Real humour is still rare in car advertising. Volkswagen, or its ad agency, DD&B (Doyle Dane Bernbach, founded in 1948), used it particularly well when it needed to fend off impending competition from the US auto industry big guns, who had the Beetle very much in their sights in 1959. The first run of press ads were disarmingly simple. Not only were there no girls, stables or country houses in the background, there was no background whatsoever. Just crisp, clean black and white photographs of the car. Designed and copywritten by Helmut Krone and Julian Koening, the first showed a tiny VW and read "Think Small". This, of course, was when most of America was thinking BIG. Big houses, big refrigerators, big TVs, big cars.

A second series of ads for the Beetle in 1960 included the brilliant "Lemon" ad. That's what it said under another undoctored photo of the Beetle. Lemon, because this is what the car might appear to be to drivers of huge, swanky Yankee gas-guzzlers of the time.

Well, how would you sell the Beetle, an air-cooled, 100km/h people's car developed in Nazi Germany to be sold for 100 Reichmarks (about £85 or $350) to the party faithful? Yet, the advertising strategy worked. Americans took the "Bug" to their auto-hearts, as they did its sporting sibling, the VW Karmann-Ghia. This pretty, but distinctly sluggish, soft-top was marketed in the States as "the world's least powerful sports car". These campaigns worked well. Sales of all cars imported to the US fell considerably between 1959 and 1963, but those of Volkswagens rose significantly.

Significant, too, is the fact that DD&B had never worked for the motor industry before. Free of the industry's tired clichés, sexism and conservatism, the agency was able to come up with something fresh, and very effective. The ads were later translated to TV. One of the most memorable shows a Beetle battling through a snowstorm. The voiceover says ,"Did you ever wonder how the man who drives the snowplough gets to the snowplough?"

Audi's enjoyable *Vorsprung durch technik* (progress through technology) campaign, dating from 1984, was in much the same tradition. It played up the way that the world sees German engineers, scientists and technicians as eccentric yet meticulous and even zealous types with clipboards and white coats. By this time, the marque was a familiar one in Britain and the US. Not long before, Audi ran a campaign, "Owdy!", to ensure that potential buyers could actually pronounce the name of the cars. Interestingly, Braun UK decided that it had all been too much to expect Brits to pronounce foreign names, and although Braun is the same as Brown, in sound and meaning, the British arm of the company decided to call itself "Brawn".

Our unintentionally funny advertising includes one of the brochures kept on my bookshelves. This one is for a finless MkV Sunbeam Alpine. I like the words "powerful" and "high performance", the blonde dolly so impressed by the hunk holding the keys to the car, the no-nonsense Bell helicopter and press cameraman in the background. In truth – I know because I once owned one – the Sunbeam, for all its gentle, cheery charms, is a bit of a lemon. It is not powerful, its performance is limited and, truth be told, it is not really a sports car.

Nor, I suppose, was the much-loved MGB, or at least not in its flabby, later days when its pretty mouth was gagged with rubber and its suspension was raised too high to take corners seriously. What could a British advertising agency in the flared and sideburned 1970s do with it? Describe it as a "mistress" while referring to its "beautiful body... a joy to handle". This was not the worst of these Carry On-meets-Penthouse-style ads. Try the 1971 ad for the Mini automatic on page 175.

Still cars and lagery, laddish ads seem destined, like Mrs Thatcher, and her legacy, to go on and on. At least Kylie Minogue, a Forces' sweetheart of our times, looks sweet and bright rather than the car industry's usual dumb and pouting stretched across the bonnet of a soft-top Ford StreetKa. Not quite Bugatti, of course, but a long(ish) way from the sexist excesses of the 1960s and 1970s.

149

150

149

1923 PEUGEOT 18CH

Ladies who shopped, if not flapped, in grand Parisian boulevards might well be captivated by the de luxe comfort of these imperious four-cylinder, 3828cc models. They were fitted with a variety of bodies, most of them opulent. Just 810 were built. The style of the advert hovers curiously between pre-and post-First World War aesthetics. The car positively screams "Twenties".

150

1921 MODEL-T FORD

Babe Ruth (1895–1948) was perhaps the greatest of all American baseball players. Born George Herman Ruth Jr in Baltimore, he was signed up early to play for the Boston Red Sox. He scored a record "hits" in his first season. Sold on to the New York Yankees, he became the kind of universally recognized sporting star and celebrity that David Beckham, the former Manchester United and now Real Madrid footballer is in the 2000s. Here's a Model-T kitted out with stylized red baseballs and big cut-out portrait of "The Great Bambino" himself. Presumably, his team was coming to play in his home town. The car was used as a promotional vehicle from its earliest days.

151

151
1935 BUGATTI TYPE 57C
This is a perfect, hand-tinted shot from a 1935 Bugatti brochure. It needs few words, and in fact there is only one – the neatly understated name of the fabulous marque. The smart young lady heading happily for the driving seat is lucky indeed. This particular model, designed largely by Jean Bugatti, short-lived son of Ettore Bugatti, the marque's founder, is just one of three Type 57Cs bodied by Atalante with a sunroof that slides down like a roll-top desk. Altogether, just 630 3.3-litre Type 57s – about 100 of them supercharged and good for 120mph – were built between 1934 and 1939. Bugatti spelt brilliance. There have been few finer cars before, then, or since.

152
1938 NASH AMBASSADOR
Nash, one of America's most successful independent car manufacturers at the time, bought Kelvinator, the refrigeration company, in 1938. This might explain Nash's pioneering development of air-conditioning in cars. The "Weather Eye conditioned air system" advertised here for the '38 Nash wasn't air-conditioning as we know it, but it was an excellent advance on the vast majority of contemporary cars worldwide that suffered badly from condensation in anything other than fine weather. It's freezing outside as the fur-wrapped doorman testifies, but it's so comfortable inside the big, eight-cylinder Nash that this elegant dame can leave her mink stole on the back of the plush seat. This advert would have had a big impact in the 1930s.

NO MORE FROZEN RIDES... OPEN WINDOWS FOR WINTER VENTILATION... CHILLING DRAFTS... FOGGED WINDOWS... DUSTY, GRIMY TRIPS... INSECTS... SMOKY AIR... OR STUFFY CARS
... the World's First Car with a "Conditioned-Air System" for winter driving is here! Always 70° comfort in zero weather. A revolutionary feature—exclusive with Nash!
WAY BELOW ZERO
And No Wrap!

152

153
1943 WILLYS JEEP
"Blazing the way in fierce counter-attacks tough, fast, hard-hitting Jeeps, manned by courageous Chinese fighting men, have inspired the recapture of many vital positions for Generalissimo Chiang Kai-shek". The Jeep was truly international. Far from peddling hype, this action-packed ad from 1943 can only just keep up with the reality of this remarkable car. It was truly one of the Allies's finest weapons during the Second World War. No wonder the Jeep, and ultimately the Land Rover and dozens of 4x4 spin-offs, have continued to woo customers, especially those who will never really need such rugged, off-road vehicles: image is all, even though, in the Jeep's case, reality matched the image. About one million were produced by Willys Overland, Ford and other approved manufacturers during the Second World War.

154
1938 PLYMOUTH ROADKING
The idea of this pre-war newspaper ad is to persuade the potential Plymouth customer that they are getting a lot of car for the money. In fact, this '38 Speedking is quite old-fashioned for its day. The best thing is to blind buyers with science. This $685 sedan is seven inches longer than one rival, and 10 inches longer than another, although we never find out what these mystery cars are. Still, we do get to learn that the Roadking benefits from "patented Floating Power engine mountings, 4-ring pistons, full-length water jackets, a Hypoid rear axle and chain-driven camshaft", which would surely have the "check books" out of our pockets before you could say Ford or Chrysler.

155
1949 FORD V8 SEDAN
Ford's immediate post-war advertising concentrates on value for money. Happy families ride out, immaculately dressed, in a car styled to look a bit like some new military jet – well, the air intake anyway – but the car is pretty lumpen and old-fashioned really. The engine was strong, but needed plenty of attention, while in the days before seat belts and crumple zones built into the structures of cars, these people would soon lose the smile on their faces if they were involved in a collision, even at low speed. But these people were just happy to be back on the road, Ford-style and in peacetime. The '49 model was largely unchanged for the next two years; it was a best-seller.

153

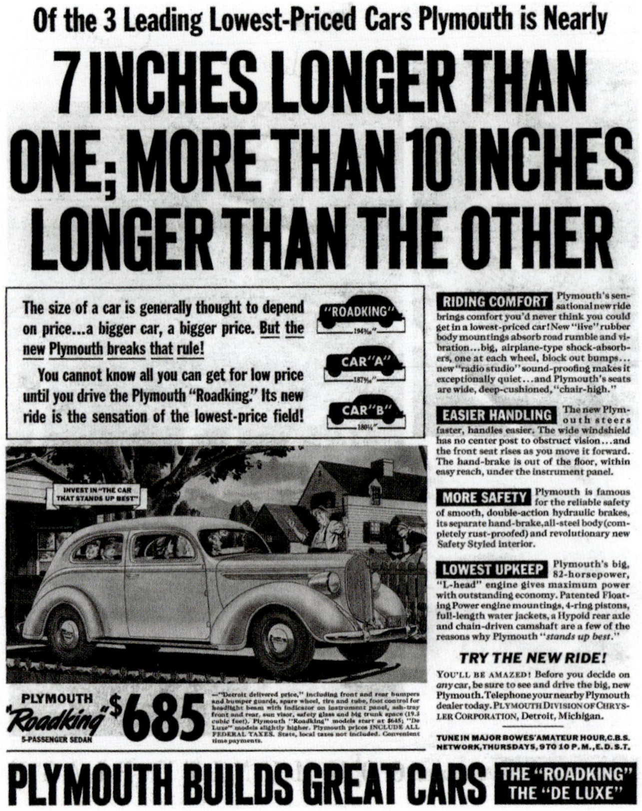

154

155

156

1958 FORD EDSEL

When Ford finally decided on the name of its big and powerful new V8 saloon, its PR director wrote a memo to company executives saying, "We have just lost 200,000 sales". Market research did indeed prove that "Edsel", the name of Henry Ford's son, was associated with "pretzel", at best, and "weasel", at worst. Still, it was safer than the names proposed, at Ford's request in 1956, by the poet Marianne Moore. She came up with, among others, Resilient Bullet, Pastelogram, Utopian Turtletop and Mongoose Civique. Perhaps she was being funny. What wasn't funny was the launch of the Edsel; it was one of the biggest and most widely advertised flops in motoring history. Maybe it was the name, or the looks, or a sudden downturn in the US economy. Whatever, the Edsel sold badly – by American and Ford standards – and this chap will singularly fail to impress the neighbours driving up in this 345hp Edsel Citation.

157
1951 BOND MK B MINICAR
Gosh, mummy and daddy, look what I've found in the woods! Crikey, Susan, it's just what we've always wanted: a super Bond Minicar! Rather! This toy-like 197cc trike is advertised here, appropriately, in the pictorial style of that ever-so-English cartoon hero, Rupert Bear, and his adventures in Nutwood (and the *Daily Express* newspaper). The Bond had first appeared in 1949, but was very slow indeed. This model, a response to customers, boasted a bigger 197cc Villiers Mk 6E single-cylinder engine, a top speed of 50mph and a fuel consumption of between 75 and 80mpg. The arrival of the BMC Mini in 1959 effectively put paid to British micro cars, although production of the Bond Minicar continued up to the 250cc, 60mph Mk G in 1966. About 26,500 Bond Minicars had been made, in Preston, Lancashire, over 15 years.

158
1957 AUSTIN A55 CAMBRIDGE
It is easy to imagine these solid English cars chugging down some warm and sunny A-road towards the summer beaches of Devon and Cornwall… forever. These were modest cars with some pretension to transatlantic styling: just look at all that jazzy chrome, that daring two-tone paintwork. Built between 1957 and 1959 – it was replaced by crisper, Pinin-Farina-designed models – the A55 could top 75mph when hurried and accelerate to 60mph from start in 31.8 seconds. This is a pre-motorway age car in the setting it was very much designed for; the British economy had recently taken off after years in the doldrums and people were dreaming of life as a prolonged summer holiday.

157

158

SEPT 1956
Announcing the arrival in this country of the famous
HEINKEL Cabin Cruiser
HEINKEL
TRAVEL IN STYLE FOR A PENNY A MILE
exceedingly
FAST
very
MOBILE
extremely
ECONOMICAL
interior spacious and
COMFORTABLE

159

160

161

159
1956 HEINKEL CABIN CRUISER
I say, is that Hermann Göring the lady is waving to? I mean, the plump chap at the wheel of the new German bubble car. Sorry, my little mistake. In fact, it's a jolly chap off with his family in his "exceedingly" fast, pear-drop-shaped car. The ad says the little car is "spacious and comfortable", although Mum's head and shoulder appear to be squished against the windscreen, or front door, or whatever you call it. This was the heyday of the European bubble car. It was not to last very long. Bet the family was pleased. The *kleine* Stuttgart *wunder* was built under licence in England, Argentina and Ireland until 1964.

160
1955 MESSERSCHMITT KR 200
A family of vampires tries to scare potential buyers from Great Britain and the Commonwealth away from the tiny, tandem-seat "Kabin Roller". Actually, the Messerschmitt was, in many ways, the best of the 1950s micro cars. It was imaginatively engineered, by Messerschmitt's Fritz Fend, well built, aerodynamically efficient, stable and fun to drive. Its 1940s Me109-style canopy was eyecatching and witty. Fend approached Willy Messerschmitt with the idea of making his earlier Flitzer mini car at the redundant aircraft works at Regensburg. Until 1956, Messerschmitt was banned from making aircraft. After discussions, the KR 200 emerged from the Messerschmitt factory. This was sold to Fend after 1956 when Messerschmitt returned to aircraft design. The KR 200's single-cylinder, fan-cooled, 191cc two-stroke engine will get it up to 60mph; steering, through an aircraft-style control column, is direct; and it will fit through gaps that a Harley-Davidson might refuse. The cars were built until 1964.

161
1957 VELAM ISETTA
A stylish Parisian lady goes to the fashion shops in her French-built Isetta. These bubble cars, nicely described by one enthusiast as a collision between a fridge, a motor scooter and an aeroplane, were first designed and made by Renzo Rivolta's Iso; Iso, a fridge manufacturer, turned to making scooters and cars in the 1950s. Then came the Isetta, which was made, under licence, in a number of countries including France. Only BMW bothered to develop the tiny four-wheeler, and this French version soon vanished from Paris boulevards, a fashion statement, here *aujourd'hui*, gone *demain*.

162

163

162
1957 RILEY 1.5

Kangol driving cap, string-backed leather driving gloves, detailed road maps, an ignition key, and off we go, Riley-style, into the A- and B-roads of 1950s Britain. The Riley was really a badge-engineered Morris Minor. Riley, like many independent British car makers, had been swallowed up by this time by the mighty British Motor Corporation (BMC). Still, the 1.5 was remarkably sporting. With a twin-carb 1,489cc engine, slick gearchange and uprated suspension, it cut quite a dash on the road. It did well in races and rallies, too. By the time production ceased in 1965, it must have seemed upright, old-fashioned and rather quaint. I had one as a runabout for five years from 1987; it was the most reliable car that I have ever owned.

163
1965 SUNBEAM ALPINE MKV

This was the last of the five series of pretty, civilized sports cars built by Rootes to the designs of, mostly, Kenneth Howes and his assistant Roy Axe. Howes had been trained as a locomotive apprentice with the Great Western Railway, Swindon, before moving on to Raymond Loewy's industrial design studios in London and New York, and from there to Studebaker and Ford. He brought American flair and comfort to this smart little sports tourer. The MkV model, the fastest and most comfortable of all, lacked the exaggerated fins of its predecessors, but was still an eye-catching design. Production ceased in 1968. Cars fitted with Ford 289 cubic inch V8s were known as Tigers. They really did roar. And go. This period brochure shows a smart Kensington gal allowing herself to be impressed by the James Bond wannabe who has just stepped out from the pilot's seat of a Bell helicopter and is being snapped by a press cameraman. The Alpine was never really quite so glamorous. But its owners could always dream.

164
1963 FORD CORTINA MK1
Chaps and chapesses off to a West End black-tie do in an East End 1200cc Cortina from Dagenham. It seems a bit unlikely. The Cortina was never – except in Lotus-Cortina guise – going to have much appeal for toffs. It was a lightweight, simple, kit-of-parts composed of earlier Ford models, but built on a generous scale and quite lively out on the motorway age roads. Although looking every inch a product of Detroit, it was named after the stylish Italian winter resort Cortina di Ampezzo, where the popular 1956 Winter Olympics had been held. The rumour was that Ford's British chairman, Patrick Hennessey, had wanted to call the new medium-sized saloon the "Caprino", but changed his mind when he learned this was Italian for "goat dung". The car was launched in September 1962. It sold very well indeed and, after its demise in 1965, it spawned three more generations of best-selling, good-value Cortinas.

165
1956 VW KARMANN-GHIA
Or the other way around, as in this German ad celebrating the car's intelligence and elegance for an elegant and intelligent audience, presumably, rather than for its power, which it most definitely lacked. This pretty, popular and long-lived car – the last were built in 1974 – had started life as a Ghia show car for Chrysler. Chrysler wasn't interested. When VW approached Karmann for a sports car, the German coachbuilder turned to Ghia, which came up with the Chrysler design… Mated to a Beetle chassis and its 1,192cc air-cooled engine, the Karmann-Ghia was born. There were many lovely ads for this car; one American TV ad showed the VW racing towards a paper screen held up by two white-coated boffins. Will it, won't it? No. The car is stopped in its tracks by the screen. "Volkswagen Karmann-Ghia", said the tongue-in-cheek commentary, "the world's least powerful sports car". It worked. What the car did have, though, was catwalk rather than assembly line looks.

164

165

166
1963 FIAT 1500 SPIDER
Quite why this svelte lady is sporting scuba-diving equipment in this blatantly sexist ad for the pretty, open-top Fiat 1500 is anyone's guess. Might there be torrential rain along the Via Appia? Is the smug dork with the Fiat's keys (yes, you the 1960s bloke supposedly gawping at the ad and dreaming of fast Italian cars and even faster Italian ladies) likely to be unable to stop before plunging into the sea at Amalfi? Who knows.

167
1975 MGB GT
But is this a saloon car or a sports car? And if this ugly, rubber-nosed version of the once pretty MGB GT is a sports car, why run away with her rather than spend a weekend with the wife. All this is nonsense, of course. As is this sensationally dumb advert from MG in the dark days of British Leyland ownership. The MGB GT was a charming sports tourer – the MGB given a Pininfarina design makeover and metal roof – not quick, but characterful. It was bought by the sort of people who tended to keep them for years, which very few mistresses are.

168
1972 MG MIDGET
She probably would, actually. The 1275cc Midget was a sweet natured and fine handling little sports car that would be just the thing for buzzing along Devon lanes in search of a slap-up cream tea with Aunty Madge and the new vicar. Here it is, posed outside every Sloane Rangers' favourite emporium, the General Trading Company, caught in a tide of entertaining clothes and trying to look groovy in Day-Glo orange paint, striped fabrics and Rostyle wheels, despite its venerable engineering.

166

167

168

169

169
1971 MINI CLUBMAN AUTOMATIC
This dizzy Goldie Hawn lookalike (one of the stars of NBC's hugely popular "Rowan and Martin's Laugh In" TV comedy, 1968–73) is doing her best to drive a Mini Clubman Automatic. Few cars could be easier, but at least she's trying. Gosh, what an amusing ad. This is the best the dreadful British Leyland, a corporation that did its level best to destroy the British car industry in the 1970s, could do. Dumb, funny only perhaps to leering salesmen in kipper ties, wide lapels and Jason King moustaches drinking lager-tops and smoking Players Number Six, this ad was enough to make anyone look abroad for their next car. Can you read the last paragraph? "It makes driving as effortless as sleeping. Sleeping, luv. You lie down, close your eyes and…" Those three dots would have meant a lot to the boys at British Leyland. Know what I mean, squire?

170
1986 AUDI 80
A very funny advert for the new Audi 80 by the London agency BBH, who had created the German marque's famous "Vorsprung durch technik" campaign. Here, the ad pretends to explain what exactly the scientific sounding phrase means. The answer is an encyclopedia's worth of Germanic engineering gobbledy-gook, ending with the sort of difficult mathematical equation that would have had Einstein scratching his head. In tiny letters in English at the bottom of the ad, it says "To fully understand Vorsprung Durch Technik you need a fine grasp of German and the mind of a German engineer. However, you can experience it by driving the new Audi 80". Clever. Very clever indeed. By the way, it means "progress through technology". More or less.

171
2003 FORD STREETKA
The neat curves of this tiny roadster based on the fish-like Ford Ka were, or so Ford's advertising team thought, neatly mirrored by the even neater curves of tiny Australian pop minstrel Kylie Minogue. So here's the Vera Lynn, the Forces' sweetheart, of our times, disporting herself and her globally famous bottom, across the pert bonnet of the imaginatively priced, fashion-conscious Ford. This is cute in a knowing, post-modern, retro-chic way, but was Ford being serious? Maybe. Kylie launched a range of underwear in London in 2003 and it sold like Model-Ts. Could Kylie pull off the same trick for the Ford StreetKa? Probably not, because the car itself was not much to write home about. Ford should be so lucky.

170

171

TRUST HOUSE
GB
EWM 25

DREAM CARS

Look at the different dreams expressed in these pages. There are cars that take their styling cues from jet fighters, some jet-powered and others that really can fly. There is even a nuclear-powered car for the nascent Space Age. More down to earth than these, although still the stuff of daydreams, are cars that promise, by way of fins and chromium plate, effortless entry into glamorous worlds; those, perhaps of svelte 1950s Park Avenue apartments, all dry Martinis, Havana cigars, long silk gloves and walls lined with paintings by Rothko.

Here are cars proffering invitations to the grand old worlds of the English country house and New England estates. Over there, cars that offer escape into altogether more exotic worlds. In between, cars whose designers, like Colin Chapman, have dreamed of creating new generations of cars, miles ahead of the mass-produced pack.

But we start this chapter with a picture that sums up one of the key car-induced dreams. That, not just of the perennially popular open-top sports car, happy young woman, fine summer's day and romantic hotel, but of escape from everyday life along the mythical "open road." The car is an MG TC, a hugely popular English sports car, built between 1945 and 1949. Many were taken back home to the US by GIs who had fallen in love with the TA and TB Midgets and wanted a new one of their own. Although neither up-to-date nor fast, the MG somehow caught the spirit of wind-in-the-hair motoring. In England it was the chosen chariot of RAF officers, a machine for sprinting, between rain showers, from Neo-Georgian officers' mess to mock-Tudor saloon bar. In the US, it made its home in sunny California or among Ivy League sophomores.

The MG TC motors delightfully through a number of films including *Love Story* (1970), *Eye of a Needle* (1981) and *Battle of Britain* (1969). In the last two, both set in 1940, it stands in for either the MG TA or TB, but only true car buffs worry about such historical niceties.

There is a lovely sense of innocence in our lead picture. Here is car with unlockable doors. It invites you, but not some hooded villain, to jump in and drive away into the blue, along highways and byways lined with foxgloves and forever June to places where there are no traffic wardens, no speed cameras, few parking restrictions and where bobbies on bicycles will only whistle and cry "blooming cheek" should you exceed 30mph past the village church.

The MG is not everyone's dream. Far from it. But it symbolizes so much of what so many of us want from the car, yet know we cannot really have it. The magnificent, supercharged Mercedes-Benz 540K promises 100mph storming along empty autobahns. The Ford Thunderbird croons of lazy, hazy, crazy days spent cruising along the Pacific State Highway. Colin Chapman's Lotus Elan and Elan 2+2 are little jet fighters for the twisting byways of Blighty. All of them sweet dreams.

These dreams start young. I remember as a boy watching one of the last Mk2 Jaguars being inched, brand new, from the doors of a London showroom. And then gunned away, gear box howling, exhaust rasping into the distance between scarlet and gold RT double-deckers and glossy black Austin FX4 cabs. That car was a 3.8-litre model on wire wheels. Its body was painted gunmetal, its interior garbed in pale blue leather. Years later, I bought a gunmetal 3.8 Mk2 with a blue leather cabin. Once restored, it was the reincarnation of the Jag I had seen so many years before. I even took it down the same street, although the showroom had become an estate agents, to connect a childhood dream with adult reality.

I remember, too, the first time I was driven by a family friend in a Mk2, marvelling at the smell of leather, the wooden dash, the deep growling purr of the engine, the rows of Second World War aircraft-style toggle switches and generous white-on-black Smith's dials. We all have our dream cars and that was mine then, even though the only Mk2 I could afford to buy at the time was made by Corgi. I still have it in a trunk somewhere.

Such dreams vary from age to age, culture to culture. Elvis dreamed of owning a pink Cadillac before he cut "That's Alright Mama", "Heartbreak Hotel" and "Blue Suede Shoes". When he had, and the royalties came pouring in, he drove back home from Nashville, Tennessee to Tupelo, Mississippi to show the folks and schoolyard friends that he had made it. His chariot: a pink Cadillac.

These dreams are shared somewhere along the line by classes, income groups and cultures. Bruce Springsteen's blue collar song "Racing in the Street" captures the mood of those down at heel but with big, open road dreams:

> I got a sixty-nine Chevy with a 396
> Fuelie heads and a Hurst on the floor
> She's waiting tonight down in the parking lot
> Outside the Seven-Eleven store
> Me and my partner Sonny built her straight
> out of scratch
> And he rides with me from town to town…
>
> Tonight tonight the strip's just right
> I wanna blow 'em off in my first heat
> Summer's here and the time is right
> We're goin' racin' in the street

Springsteen has also sung of pink Cadillacs, beaten-up Buicks and so many cars and journeys that capture the heart of the great American dream of just getting up, turning the key and going, someplace, who knows where, possibly, like Thelma and Louise, where the road runs out.

Designers continue to produce weird and wonderful "concept" cars, and they should be applauded, for although we know that the car is a form of collective madness, it is also part of a collective dream. In Ghana, wonderful coffins are made in the guise of the deceased's favourite Mercedes-Benz. The late Sri Lankan architect, Geoffrey Bawa, kept his favourite white Rolls-Royce Phantom III parked in his exquisite Colombo living room. When one-time publisher of the *Architectural Review* Hubert de Cronin Hastings could no longer drive his Phantom III, for whatever reason, he had the magnificent V12 machine lifted on to blocks and used it as a study in the grounds of his Sussex farm.

So many cars crowding our streets seem lacklustre and prosaic; perhaps, though, each dreams of being an MG TC, T-Bird, Phantom III or pink Cadillac.

172

173

172
1935 HUDSON CUSTOM EIGHT
Just what is he saying, and why is she smiling? Who knows? Whatever these two are up to, they've arrived in a very dashing Hudson. It's a lovely summer's day and there must surely be a handsome wicker hamper and a chilled bottle of champagne in the trunk of the semi-streamlined sedan. This is an enduring dream; the car as mechanical Cupid.

173
1945 RILEY RMA 1.5-LITRE
A very British, austerity era dream… The rakish, fabric-roofed, two-tone Riley is nowhere near as fast as it looks. With just 55hp under its dashing bonnet, this car will be pressed to beat 75mph. But it cuts a lovely dash, and the girls will soon be on their refined, twin-cam way from the airfield to scones and cream in some chintzy cheery Sussex tea shop, I'll be bound.

174

174
1936 MERCEDES-BENZ
540K CABRIOLET

The mighty 500/540K cabriolets were never really the Nazi monsters that they have been labelled. Designed by Hans Nibel, they were truly magnificent machines designed to cruise at very high speeds in their high fourth gears along Fritz Todt's new autobahns. But they were not just fast; they were sophisticated, too. The 112mph supercharged 540K featured power-assisted brakes and independent suspension, front and rear. Bodywork varied from the gross to the voluptuous. They remain completely in their element along sweeping autobahns, dream cars that have long outlived the Nazi nightmare.

175
1937 MERCEDES-BENZ
540K SPECIAL ROADSTER

Hermann Göring's favourite car. Even so, a truly beautiful mechancial Valkyrie, and almost a parody of the dream, open-topped roadster. That impossibly long bonnet, pert rump, sweeping running boards, chromed exhaust pipes, tiny windscreen and just the two, well-cushioned leather seats. And does it go? Exactly as it looks. And as it should with a 180hp, 5401cc supercharged straight eight. Just 26 of these three-ton Wagnerian supercars were built, with bodies by MB's own *karoserrie* in Sindelfingen. They cost 28,000 Reichsmarks (28 times the price of a Volkswagen) and 40 per cent more than the most expensive V16 Cadillac.

175

176

177

176
1934 CHRYSLER AIRFLOW
Here tradition means modernity as a cluster of highly posed New England huntsmen stop to take a gander at this latest streamlined wonder. A brave stylistic move by Walter Chrysler, the Airflow was the automotive equivalent of William Van Alen's iconic Chrysler Building scraping the sky in mid-town Manhattan. Sadly, the public found it hard to take to the car's looks, especially its waterfall grille. It also suffered more than its fair share of teething problems. And yet the Airflow, developed by Carl Breer with advice from Orville Wright, and with the aid of a new wind tunnel at the Chrysler works, led the way to the architecture of the modern car. A flawed dream, but one that was to change the way cars looked worldwide.

177
1940 FORD V8 CONVERTIBLE
Okay, you couldn't afford a drop-top V16 Cadillac; well, what about this for just $849? With a sturdy V8 engine under the hood and snazzy looks, this 1940 Ford offered cheap(ish) and glamorous motoring to people like these in the period advert, who wanted the good life but had average incomes.

178

178
1923, LEY STREAMLINER
This might have been the democratic Weimar Republic Peoples' Car a decade before Adolf Hitler came to power. These odd looking cars running down Unter den Linden are three experimental, tear-shaped streamliners designed by the Hungarian engineer and aerodynamicist, Paul Jaray (1889–1974). They were never put into production. Jaray went on to work on the design of the Tatra 77, and influenced Ferdinand Porsche's work on the Volkswagen.

179
1945 ZIS 110 LIMOUSINE
The Soviet proletariat could only dream. The mighty seven-seater Zis was never put on sale. It was reserved for the Communist Party elite. It was also, rather engagingly, a handsome rehash of the 1942 Packard Super Eight. One of these plutocratic capitalist cars had been presented to Stalin by President Roosevelt in 1942 as a goodwill gesture. Powered by a mighty 6005cc straight eight, the three-ton Zis could sweep up to 140km/h on Moscow's vast roads, passengers hidden behind plush velvet curtains and armour plate.

179

180

181

180
1959 PLYMOUTH SPORT FURY CONVERTIBLE
One way of spending Sunday afternoon with a long-necked lady or two. What on earth did the lady in the Chinese hat think of Bud's decision to take her for a spin in a zoo? How did they end up in the giraffe enclosure? Did the power brakes on the svelte new Sport Fury fail? Unlikely. These were cars were tough as well as chic. Powered by a 318 cubic inch, 230hp V8, they were also powerful beasts. Was this ad meant to prove that the '59 Plymouth was as elegant, exotic and as long legged as a giraffe? Whatever, it evokes the idea of escapism on wide open freeways that, with the right car, might just take you anywhere.

181
1955 FORD THUNDERBIRD
American soldiers coming home from Europe between 1945 and 1955 brought many sports cars back with them. But where was the affordable all-American sports car? It remained a dream until the launch of two legendary rivals, the glass-fibre-bodied Chevrolet Corvette in 1953 and the all-steel Ford Thunderbird two years later. The T-Bird offered American levels of comfort and even automatic transmission, but with real style and performance. It could top 115mph and cover the standing quarter mile in 17 seconds. Jaguar XKs were faster, but T-Birds were relaxed cruisers as well as pretty quick. Named after the New Mexican thunder god, and styled by Frank Hershey and William P Boyer, the Thunderbird was to become one of the enduring icons of American design.

182
1959 CADILLAC SEDAN DE VILLE
No wonder she looks happy. This is the ultimate all-American automobile dream. The '59 Caddie: 225 inches of swooping curves, flying fins, jet fighter details, acres of chrome, air suspension, cruise control and power everything – brakes, seats, even the door locks. And powerful, continent-crossing performance promised by a 390 cubic inch, 325hp V8. All this for $5,080. This tour-de-force, styled by a team led by Cadillac's Ed Glowacke, was an attempt to outdo Virgil Exner's contemporary designs for Chrysler. Many critics thought Cadillac had gone too far in 1959, and the next year's models were tame by comparison. Today, the '59 Caddies are very much collectors' items. The only one I've driven has been in Communist Cuba. They love them there.

183
1959 CHEVROLET IMPALA SPORTS SEDAN
I had a pink Corgi model of this car when I was young and it was long one of my favourites, even though I knew that I really preferred Aston Martins and Jaguars. There is something more than just over-the-top in these flowing lines; the car, one of the last designs supervised by General Motor's long-term head of styling, Harley Earl, is almost glorious. Even the most hard-bitten anti-American revolutionaries took to them: Che Guevara drove away from his second wedding in one.

182

183

184

185

186

187

184
1951, GENERAL MOTORS LE SABRE
With its jet-fighter-style nose cone and air intakes, wrap-around windscreen and tailfins, Harley Earl's Le Sabre concept car stunned audiences at the 38th Paris Automobile Show. Made of sheet aluminium and magnesium and powered by a supercharged V8 sucking in methyl-alcohol as well as petrol, the Le Sabre was fully intended to be a jet aircraft on the road. If it started to rain, the roof would rise automatically. Flat tyre? The car could jack itself up for you. Among those taken on a wonder ride in this testbed of practical tricks were Dwight D Eisenhower, King Hussein of Jordan and Bob Hope; but it was a dream, not a joke.

185
1958 FORD NUCLEON
It came from Ford's Advance Styling Studio. It was a one-third scale model and nothing more, otherwise we might not be here to tell the tale. The Nucleon was designed to be powered by a small nuclear reactor inset between its pronounced rear fins. If you could live with this idea, then just look at the way the wheels are placed and the prognathous overhang at the front of the Space Age beast. It would have handled like an oil tanker in a hurricane. With enough power to blow up the whole of Detroit behind your head, you would have driven this car of the future very slowly indeed. Some dreams are best left dormant. This is one of them.

186
1956 OLDSMOBILE GOLDEN ROCKET
This metallic bronze sci-fi car is hanging out on the beach and soaking up the sun. A delightful mid-1950s dream, it was built of glass-fibre to keep weight down and with 275hp under its jet-style bonnet it ought to have been pretty quick. But whoever knows with concept cars? They are designed partly to test new ideas and components, and partly to grab media attention. Getting inside was easy for the laziest consumer-era driver and passenger: raise the gull-wing doors and, hey presto, the seats swivelled out while the steering wheel tilted away. A car for truly lazy, hazy, crazy days in summer.

187
1956 GENERAL MOTORS FIREBIRD II
Harley Earl's gas-turbine-powered dream car inspired by the Douglas Skyray jet fighter. The car was designed to be driven by a "computer"; it would "steer itself". And, on a special test track, the General Motors' research team just about got this real-life Batmobile to do what Harley Earl wanted it do. It was never exactly practical, yet it still survives and has encouraged American car designers to think hard about the next generation of truly automatic, drive-by-wire, hydrogen-powered fuel-cell cars.

188
1959 JAGUAR MK2
Set in an idyllic English Cotswold country house scene, here is the quintessence of English saloon car design in one of its natural habitats. The house is actually a bit of an upstart, a twentieth century take on an old Cotswold home, just as the dynamic Jaguar was a bit of an upstart, too. Not quite a Bentley, and a little uncouth when pushed hard. Never mind, Jaguar's William Lyons knew exactly how to realize grand motoring dreams for relatively little money. The 3.8-litre Mk2 was the fastest four-door saloon car in the world when launched in 1959: 125mph and acceleration from rest to 100mph in 25 seconds was very much a reality.

189
1968 LOTUS ELAN S4 AND LOTUS ELAN 2 + 2
Here is Colin Chapman (1928–82), one of the great British car designers, outside his home with two of his nimble sports cars. Chapman, a one-time RAF flying officer, played an enormous role in the development of Formula One racing. He also designed and built a range of very successful, hugely enjoyable lightweight sports cars, including this pair of short- and long-wheelbase Elans. Each Chapman Lotus was pared to its alloy bones. They remain thrilling to drive today, truly modern cars that played to Ludwig Mies van der Rohe's famous architectural maxim: less is more. Strange, then, that Chapman's house seems so very old-fashioned and unstylish – my dears, just look at those net curtains – but that's England for you. We dream of compromise in all things. And even the Elan was forced to live with a wood-veneered dashboard.

188

189

190

191

192

190
1963 ASTON MARTIN DB4
Styled by Touring of Milan, with a chassis by Harold Beach and powered by Tadek Marek's magnificent 3.7-litre, twin-overhead-cam straight six, this is one of the great cars. Nothing in excess. Simple fighter-style cabin. No wood. Little in the way of chrome. Lightweight body construction. 140mph. Launched to critical acclaim in 1958. Replaced by the *Goldfinger*-style DB5 in 1964. Only bettered by the faster, short-wheelbase DB4 GT.

191
1946 CISITALIA 202
The dream of a former footballer and amateur racing driver turned business magnate, Piero Dusio, the Cisitalia was styled by Pininfarina and Vignale and was one of the most stylishly modern cars of its time. It must have seemed a dream when first displayed to the public at the Italian Grand Prix in 1947. The car had a racing spaceframe chassis, and making liberal use of components from the workaday Fiat 1100, it was a perfectly practical proposition. Performance was lively from the 60hp 1089cc Fiat engine because the car was so light. Its forward-looking styling earned it a place in the New York Museum of Modern Art's permanent design collection.

192
1965 ASTON MARTIN DB5
With a few slight detailed changes, the Aston Martin was transformed into the lithe 282bhp DB5. This was the car chosen for James Bond in *Goldfinger* (Guy Hamilton, 1964) as modified by production designer Ken Adam with ejector seat, machine guns and other cunning devices. The film was hugely successful; so was the car. It was very many little boys' dream car, and doubtless, little girls', too.

193
1973 FORD MUSTANG
A 1970s American dream summed up in one simple snap. Sunshine, a pretty girl in groovy outfit at once body-hugging, and flared like the car, and a Mustang to get away in.

193

194

194
1949 AEROCAR
Yes, it really can fly. Here is a dream come true, but one that never took off in commercial terms. Perhaps this is just as well. Can you imagine the sky full of pilots who have difficulty driving a stock Chevrolet? Mott Taylor's Lycoming-powered Aerocar remains a wonder. This is no nutty professor's daydream and the engineering is thorough. The glass-fibre-bodied car, which can run on the road at up to 67mph, can be converted into a plane in just five minutes. The car carries it wings folded up behind it on a trailer. Wings in place, up she goes, able to cruise at 100mph-plus for 300 miles. As late as 1970, Ford expressed an interest in mass-producing an aerocar, but nothing happened. Just as 99.99 per cent of cars will never fly, except accidentally, we are not all born to be pilots.

195
1988 FLYING CAR
Here's another way. Stick wings on your Ford Capri, and it's up, up and away.

196
1999 MOLLER M200X
Flying saucers are usually only ever seen in the United States. This is either because aliens prefer to take their holiday there rather than in Uzbekhistan or because they have unwittingly witnessed flights by Dr Paul Moller's M200X flying saucer car. This first flew in 1989, and has since made more than 200 short trips. Today, Moller International is hoping to have its vertical take-off M400 Skycar, a *Star Wars* fighter lookalike, on the market for about $600,000 in 2006. Commuting will never be the same.

195

196

197

198

197 + 198
1936 STOUT SCARAB
Only six or nine – no one seems sure – of these streamlined, rear-engined, aluminium-bodied cars were built between 1936 and 1939. They were remarkable machines with generous interiors that could be re-arranged ingeniously. The engine was a reliable 100hp Ford flat-head 221 cubic inch V8. The Scarab boasted all-round coil-spring independent suspension, and handled pretty well. It was designed by William B Stout (1880–1956), inventor, aviator and one-time editor of *Motor Age*. It must have looked very odd at the time – an American equivalent of the Volkswagen, although the two machines were developed entirely independently – and failed to sell. Shame.

199
1934 DYMAXION III
Designed by the American engineer, inventor, environmentalist and sage, Buckminster Fuller (1895–1983) wanted to mass-produce a car using ideas, materials and technology from the aircraft industry. The result was the sci-fi-style Dymaxion. These three prototype aluminium-bodied 10-seaters were powered by a rear-mounted Ford V8 that drove the two front wheels, while the single rear wheel steered. "Bucky" made all sorts of fantastic claims for the Dymaxion - it could accelerate to 60mph in three seconds, it had a top speed of 120mph – but although it rode well in a straight line, the Dymaxion shied away from tricky things, like bends. An accident involving the prototype and the death of an enthusiastic English visitor ensured that Fuller never got the funding he needed to develop this glorious failure. He did, though, go on to invent the geodesic dome and remains a hero among architects, designers and imaginative engineers.

199

200

200

1958 FORD EDSEL CITATION

It was meant to be the big V8 Ford that everyone wanted. Market research had proved it before the car had turned a wheel. The Edsel was one of Ford's rare flops, and a big one: it sold fewer than half the numbers expected. What the styling department in Detroit was on when they shaped that infamous grille, we will never know. Still, Edsel himself knew more than a little about excess. He asked Albert Kahn, the architect of Ford's Detroit factories, to design him a lakeside home in the English Cotswolds style. It added up to a 60-room mansion stuffed with English and French antiques, and several rooms filled with works by Matisse, Cezanne, Franz Hals and Diego Rivera. Since Edsel died in 1943, there would have been no '58 Citation in his garage. The house, in Macomb County, Michigan is open to the public.

201

1975 VANDEN PLAS 1500

A posh limousine, a miniature Austin Princess for everyman, this is the Vanden Plas 1500, one of the most delightfully pretentious cars of all time. Based on the curious, boiled-sweet shaped Austin Allegro, the 1500cc front-wheel-drive saloon boasted a luxurious leather and walnut-veneered interior and a Bentley-style grille. Originally a Belgian supplier to the early motor industry and then a coachbuilder to such illustrious marques as Rolls-Royce, Alvis and Lagonda, the British arm of Vanden Plas, was founded in 1913 and became a subsidiary of Austin in 1946. The Kingsbury works in north London closed in1979 and Vanden Plas became nothing more than a smart name. The 1500 model is much prized in Japan as an example of English eccentricity. The first popular model to get the Vanden Plas treatment was an 1963 Issigonis-designed Austin 1100 done up in leather and walnut for Fred Connolly, leather supplier to the motor industry. It was much admired – so much so that Austin put it into production. The Japanese collect this endearing model, too.

202

1975 AMC PACER

Long before most mass-production cars began to look like half-sucked boiled sweets, Richard A Teague (1923–91) designed the blobby Pacer for AMC (swallowed up by Chrysler in 1987). It was wide, spacious and not as bad as it was made out to be by those who found its shape hard to swallow. Just 280,000 sold between 1975 and 1980 – chicken-feed in the US – and few survive. Originally it was to have included several advanced engineering features, including a rotary Wankel engine, but in the end it was pretty conventional. It looked, said too many for the car's good, like a frog or a pregnant guppy.

201

202

Parking
Tickets
SHORT TERM
PAY CAR PARK

CARS AND SEX

Sex has been used to sell cars for at least a century. Here, though, we open with a strappy-heeled lady, nicely wrapped up in fetching belted raincoat and gloves, circa… well, guessing by the Austin Allegro and Renault 6 in the background… 1973, and close with the open-legged Page Three model Jordan astride the penile nose of a Formula One Jordan, a quarter of a century later. *Plus ça change, plus sont les choses differents*. Or lack of them.

Sandwiched between our English models are leggy ladies servicing cars in a variety of ways. Washing and waxing, filling and jacking up. Most of these images are either hilariously or horribly sexist; but, this is, I suppose, the point. Cars were sold by the motor industry as accessories for men, which, in most of these images, the bikini-clad women are, too.

There are, though, a few gloriously camp and evenly genuinely sexy images here which show that things did not always have to be this way. I love the photographs of Lady Docker, the 1950s Daimler queen, decked out in an extravagant frock embroidered with swirly Ds and Daimlers; and, again, her ladyship powdering her nose in the back of a sensationally kitsch Docker Daimler. This one is her 1953 Daimler Silver Fish decked out in red crocodile skin. Nice.

Lady Docker had been Norah Collins, a one-time dancer at the Café de Paris in Leicester Square, before she married Sir Bernard Docker, chairman of Daimler long before this venerable company was swallowed up by Sir William Lyons's Jaguar. She was a charming, eccentric and popular gal. She once invited 40 miners aboard her private yacht, *Semara*, and served them pink champagne. When £15,000 of her jewels were stolen, Billy Hill, self-styled "King of the Underworld", vowed to get them back. "I admire Lady Docker," he said. "She's not afraid to mix with people like me." Sir Bernard lost his job at Daimler in 1956, and the Dockers's luck ran out. They ended up living in a bungalow in Jersey, a couple who had added immensely to the gaiety of the nation during the glum years of po-faced Austerity.

Marilyn Monroe looks glum in one of the pictures on page 224. Happy or sad, this screen goddess always knew how to play for the cameras. Here she signals a brace of poignant farewells from the windows of big Yankee cars.

Bored, not glum, is the word to describe the look on the blonde dolly bird's face on page 228. This advert dates from 1968. It shows a scrum of young men clearly besotted with the thrilling 1296cc engine of a lipstick red Triumph Spitfire while oblivious to the charms of our standard-issue 1960s sexy blonde in tight fitting top and mini-skirt. The contrasting image below, from the late 1990s depicts a pair of matching blondes in mini-skirts, again, giving one of the last new Issigonis Minis the once over. This time, though, the girls look confident and happy. They clearly have careers. They are not waiting around for boorish men. And the bonnet of the bright red Mini is firmly closed. It goes. Who cares about its A-series engine? It is not a sexist or sexy car. It's more like a puppy dog or cuddly toy.

Some cars, though, are obviously sexy beasts. Lamborghini Miura. De Tomaso Mangusta. Chevrolet Corvette Stingray. Ford GT40. Citroën SM. Some are lithe. Series I E-Type Jaguar. Fiat Dino Spider. Lotus Elite. Others are louche in a "come up and see me sometime" way. Facel Vegas. Bentley Continentals. Lincoln Continentals. But no cars really live up to the curious sexual attributes afforded them by all too many magazine and newspaper writers over the years. A car is a machine, however gorgeous its apparel, curvaceous its body, and no matter whether it hugs its driver, cossets passengers or purrs sweetly into their ears.

That the car can be an extension of someone's sexual ego, whether inflated or deflated, is another thing. It has long been said that cars with long bonnets, like the much maligned E-Type, are penis extensions, or substitutes, and perhaps some men do think this. Male sexuality has a lot to do with projecting and projections of one sort or another. But this is cod psychology. Women are attracted to E-Types, too. Oh dear, this is getting sticky and likely to end, if not in tears, then in dreadful puns and Carry On style double entendres. Or else stuck up a cul-de-sac without reverse gear, missus.

It does seem odd, though, in an era – ours – in which gross sexual imagery not so much abounds but is all but mandatory that the motor industry continues to rely so heavily on old-fashioned sexual values. Motor shows worldwide still feature dolly birds preening and pouting across the bonnets of very ordinary cars, as if somehow their polyester-nylon charms will persuade a sex-starved clientele to part with well-earned wages for a blob of uninteresting plastics and metal. Or are they there for bored sales reps stuck on show stands for days on end to gawp at, a kind of live tabloid tableau?

This thinking – or gutting, more like – seems to be a 1950s hangover. Look at some of the pictures in this chapter. Were British motorists really persuaded to fill up at a petrol station where goose-bumped Barbara Windsor lookalikes in bikinis and stilettos manned, as it were, the pumps? I suppose I might just be taken in by the charming lady in a nice frock offering a ride in her Bond Minicar on page 219. Ever so smart, she likes to drive indoors on swirly patterned carpets between classical columns. Always smiling. Very courteous. But she's not exactly Jordan, innit, I mean, is she? In fact, I should imagine she is called Janet or Jennifer. A good sort in a Miss Joan Hunter Dunn way, but, sadly, not sufficiently sexy enough for our cynical, thrusting, pouting, ruthlessly commercial culture.

Cars and sex go together in rather different ways than shown in these advertisements, shows and gimmicks. Cars offer freedom to lovers, young and old, to get away, places to kiss, and more. In this sense, some of the cars that magazines tend to describe lasciviously are some of the least sexy of all. Have you ever tried to kiss in a Ferrari or Lamborghini? Not only are you trapped in bear-hugging seats, but there is a vast central gearbox tunnel between you and your beloved. Your legs, meanwhile, are squeezed into a gap that can just accommodate brake, throttle and clutch. You would do more for your sex life in a Humber Hawk.

203

203
1917 OLDSMOBILE
The world is at war, but you would never guess from this charming portrait of an American beauty modelling a white silk dress on the running board of a gleaming '17 Oldsmobile.

204
1955 FIAT 1400 SALOON AND 1100 103TV TRASFORMABILE
Most people wear blue overalls at oil refineries, but not this Italian model who has chosen a tightly fitted coat and fur stole – on a sunny day, too. This fashion shot was taken in Milan in 1955. The cars are brand new, although the pretty 50hp Trasformabile, with its shark-nose grille and wrap-round windscreen, looks decidedly more modern than the stolid, taxi-style 1400 alongside. Cars were much used in fashion magazines in the 1950s and 1960s.

204

PIRELLI

LOMBARD
LIMITED
LOMBARD HOUSE CURZON STREET PARK LANE LONDON W. 1
A.T.SMITH
LOOSE COVERS FOR CAR SEATS
Goggomobil

205

205
1956 GOGGOMOBIL T300
Go on, guess who the leopard skin model is… Jackie Collins, author of *The Stud* (1969), *The Bitch* (1979) and *Deadly Embrace* (2002) no less. While glamorous Jackie has gone on to sell more than 200 million books, the Goggomobil has fared less well. In fact, the Hans Glas's car company was taken over by BMW in 1969. Jackie is sitting in what appears to be a special luxury edition of the 15hp, 293cc 95km/h German saloon. She will move on to better things 12 years later when she publishes her first best-seller *The World is Full of Married Men* (1968), but few Goggomobiles.

206
1957 SIMCA ARONDE
There we were, Monsieur le Gendarme, just driving back from the ball and suddenly half the *voiture*, how you say, vanished. A strange episode from the 1957 Paris motor show featuring two belles in extravagant frocks and half a car. Would such grande dames really have chosen a humble Simca?

207
1964 ASTON MARTIN DB5
Here's some real English-style glamour: leggy bottle blonde, gloriously seedy backdrop – ladder, plank, assorted junk – with one of the sexiest cars ever looking good despite every attempt to reduce it to the level of a prop in a *Carry On* film. The DB5 remains one of the most coveted of all British cars.

206

207

208

208
1966 FIAT BERTONE SHOW CAR
Slight lapse of taste here on the Bertone show stand. The model looks decidedly uncomfortable in a huge, busby-like wig and bizarrely engineered bikini. Still, she hides the car, which is clearly not one of the great carrozeria's best.

209
1969 GP RACER
'Allo, dahlin'. Sheer, sex-sational class, innit, at the British Grand Prix, 1969. A fast, sticky lady, perhaps, but not quite so fast as the slippery Jackie Stewart, who won the race in a Matra MS80.

209

210

211

212

210
1968 FIAT 124 SPIDER
Car turns into model who turns into car. Peggy Walton, a model at the 1968 New York motor show, seems all but indistinguishable from the pretty Fiat Spider. A funny, and almost innocent, image.

211
1987 TVR-ES
Fast, eye-catching, blowsy and loud, TVRs are built in Blackpool, Lancashire. Here car and motor show models are in perfect accord.

212
1988 PANTHER KALLISTA
Tasteful, or what?

213
1999 TOYOTA AXV CONCEPT CAR
Sometimes, we are all stumped for words. This is very possibly Toyota's homage to the Spice Girls. Or not. Tokyo motor show, 1999.

213

214
1955 BMW ISETTA
Hopping one-legged style from a German bubble car. It might have been a fashionable pose, or was it just cramp? Natty outfit matched by the car's stylish tartan upholstery.

215
1960 TOYOTA TIARA
Model Diane Chiljan jacks up one of the new 75hp Nipponse sedans aimed squarely at the US market; she splits her skirt in the process. Women and cars, I ask you...

216
1956 BOND TYPE-D MINICAR
Here's lovely would-be debutante Gay McGregor reversing the 197cc '56 Bond across a swirly carpet at the three-wheeler's debut at a London hotel. That year's Suez Crisis, when Britain unwisely attempted to invade Egypt, led to a steep rise in the price of petrol. A parsimonious Bond made some sense. Bet the charming Miss McGregor with her Sussex teashop smile didn't own one, though.

214

215

216

217

217
1963 HILLMAN MINX CONVERTIBLE
Fetching sleeveless dress in pink or blue and white checked cotton with collar and pussy-cat bow in white organdie. By Bernshaw. Modelled by fetching gal smoking a fag in squishy 1,494cc Minx.

218
1969 US MUSCLE CAR
"A driving outfit designed for long or short trips in the summer season." Flares are back, in 2003, but whatever happened to driving outfits?

218

219

220

219
1955 FORD POPULAR
Blimey, mush, cop a load of that, and no mistake! Cockney geezer, circa 1960, gets pumped with gallons of free petrol from a Barbara-Windsor-style dolly bird. This image is either very funny or rather sad. Still, the MoT test is only a year away at best and the Popular may well be off the road for good, while our bikini clad lovely might be able to buy some clothes before she catches a death as the British economy picks up.

220
1954 FORD CONSUL MK1
Almost impossibly English scene. Charming schoolgirls in bikinis – Stella Long, Marilyn Woolhead and Karen Lewis – buff, polish and prod some battered and wonderfully dumpy gor'blimey Ford at a petrol station at Ealing, west London as part of a publicity campaign by English Petroleum in November 1965. That's right, November, a good month in England to prance about in bathing costumes.

221
1954 DAIMLERS
Here's Lady Norah Docker in a modest and tasteful white satin number embroidered with sequins depicting hubby's Daimler cars. The occasion is a fancy dress ball at the Poole Harbour Yacht Club in January 1954. The policeman is Lady Norah's son, Lance.

222
1954 DAIMLER STARDUST
M'lady Docker in furs, attracting press attention in the plush salon of her '54 Daimler. Rationing was just coming to an end in Austerity-era post-war Britain when this fetching snap was taken. Although extravagant and decidely camp, the former Norah Collins – a one-time dancer at London's Café de Paris – was a popular figure during the years her husband, Bernard Docker, was chairman of Daimler. She added greatly to the gaiety of the nation in glum, monochrome times.

221

222

223
1954 JOE DIMAGGIO'S CAR
Marilyn Monroe leaves her Beverly Hills home in tears, with baseball star husband Joe Dimaggio at the vast wheel of their slinky car. Marilyn was about to sue for divorce. She knew how to look good for the cameras even in times of distress.

224
c 1960 MARILYN MONROE LIMOUSINE
Making a cute and knowing automotive exit with next husband, Arthur Miller.

225
1956 FORD THUNDERBIRD
American dream. Marilyn Monroe and Arthur Miller setting off for a picnic from their Roxbury, Connecticut home on June 30, 1956, the day after their wedding. Marilyn starred with Don Murray in Joshua Logan's *Bus Stop* that year. Miller had achieved international fame as a playwright with his *Death of a Saleman* (1949). The happy couple divorced in 1961, the year John Huston's *Misfits* starred Monroe at perhaps her very best and with a taut screenplay by Miller.

223

224

226

227

228

226
1962 RENAULT CARAVELLE S
The divine Brigitte Bardot and two pets in St Tropez in 1962: a dachshund and a rear-engined, 956cc Renault Caravelle S. The Renault was a bit like the VW Karmann-Ghia: not fast, but lithe and gently voluptuous. BB, however, was definitely more than a match for both. She had promoted Renault cars from 1960.

227
1955 SUNBEAM ALPINE TOURER
Here's Grace Kelly with Cary Grant on location on the French Riviera during the filming of *To Catch a Thief* (Alfred Hitchcock, 1955). The British director said she had "sexual elegance", and who would say Hitch was wrong? This was when the beautiful actress met her future husband, Prince Rainier III, at the Cannes Film Festival. She became Princess Grace of Monaco the following year. She died in Monte Carlo on September 14, 1982 at the wheel of her Rover 3500. She was said to have suffered a stroke; the car plunged 45 feet from a twisting cliff road.

228
1959 CADILLAC FLEETWOOD
This windy picture is from Gina Lollobrigida's family album, dated January 1, 1970. Nice to see a '59 Caddie getting the glamorous star treatment years on from its sensational debut. "La Lollo" was one of European cinema's first post-war sex symbols. She starred in dozens of films in both Italy and the US in the 1950s and 1960s. She is also a talented photo-journalist as *Italia Mia* (1973), her first book of photographs, amply proved.

229

230

229
1968 TRIUMPH SPITFIRE
I say, look at that cracking 1,296cc engine, chaps! Much bigger than the previous 1,147cc unit, eh? Standard-issue British men ignore standard-issue British blonde.

230
1998 MINI
Cheery mini-skirted blondes, free from the attention, or lack of it, of standard-issue British men, click-clack past one of the very last new Issigonis Minis. The Mini was a mechanical cherub, cheeky if never exactly sexy.

231
1998 JORDAN GP RACER
Jordan, the saucy 34FF tabloid model astride the priapic nose of a Jordan at the 1998 Spanish Grand Prix. The team did well that year, with Damon Hill and Ralf Schumacher coming home 1-2 at the Belgian Grand Prix. Jordan – Katie Price – went on to much bigger things, standing for parliament as candidate for the Greater Manchester constituency of Stretford and Urmiston on a "free plastic surgery for all ticket". She didn't win, but continued to wow paparazzi, footballers, sticky schoolboys and the 'ello darlin' British tabloid press.

231

CD
75-IT5119
F

for the Stars Organisation for Cerebral Palsy. Here is a much-loved star with more than a mechanical heart.

Ken Adam, himself, has owned several glamorous cars. These include a Mercedes-Benz 540K, an E-Type that he drove director Stanley Kurbrick to and from London and Elstree Studios each day during the filming of *Dr Strangelove* at no more than 30mph ("otherwise Stanley, who hated speed, would threaten to get out and walk"), and for many years since then a rare, short-chassis Rolls-Royce Silver Cloud cabriolet.

Other cars can dive under water – James Bond's Lotus Esprit in *The Spy Who Loved Me* (Robert Zemeckis, 1977) – or even travel back in time as the gull-winged De Lorean does in *Back to the Future* (Robert Zemeckis, 1985).

In real life, TV and cinema stars have often coveted the cars they have driven on screen. Roger Moore, who starred as The Saint on the small screen before taking on the role of James Bond on the big screen, drove a P1800 Volvo Coupe just as his fictional character did. Other actors progressed to cars that were perhaps too glamorous, or too expensive, for use on film sets. Gary Cooper and Clark Gable owned the only two short-chassis, supercharged Duesenberg SSJs, two of the greatest of all American cars.

Cars are used whenever a bit of excitement is needed or a scene needs setting. Or they reinforce the view we have of the role actors are playing. In *Diva* (Jean-Jacques Beneix, 1982), for example, the conceptual artist hero drives a glossy black Citroën Light Fifteen Traction Avant. When villains blow it up and assume he is killed in the explosion, he opens a garage door and reveals a second identical car. Just as the car is very cool, so is the character; so, too, this sequence in a film which must have done wonders for sales of old Citroëns.

Films have this power. *Goldfinger* (Guy Hamilton, 1964) did wonders for sales of Aston Martin DB5s; this svelte, iron-fist-in-velvet-glove grand tourer – in other words, Bond-on-wheels – has been associated with the film ever since. Patrick McGoohan's cult TV series *The Prisoner* (1967) must have helped sales of the Lotus Super Seven, a car that screamed rebellious individuality. The jolly British comedy *Genevieve* (Henry Cornelius, 1953) was pivotal in the fortunes of the classic car movement. Hugely popular worldwide, the film starred Genevieve, a 1904 10/12hp Darracq braving the London to Brighton run. Before the film, old cars were considered either (a) scrap, or (b) playthings for eccentrics. After *Genevieve*, classic cars became desirable and ultimately fashionable.

Pop stars have done much the same for various makes and models too, although few people could ever afford the expensive models they chose to flaunt, and then, like The Beatles, decorate them from bumper to bumper in psychedelic paint finishes.

The car grew up with the movie business. The Lumiere brothers gave the world the moving picture in 1895 and Carl Benz the basis of the car some 10 years earlier. Rock 'n' roll and the car have often been synonymous, the inspiration for the soundtracks of countless millions of journeys on and off screen. The car has always had stars in its headlamps. You will have your favourites. I hope some of them are here.

METRO-GOLDWYN-MAYER-ST

233

234

235

232
1959 MERCEDES-BENZ 190SL
Alfred Hitchcock, at the height of his powers, and at the wheel of his Merc leaving the MGM studios after a day's work on his classic thriller *North by North West* (1959), starring Cary Grant and James Mason.

233
1959 MERCEDES-BENZ 190SL
Here's Yul Brynner, in 1959, off for a spin in his 190SL, clearly very fashionable that year in Hollywood. No time to change out of his costume; he was starring in *Solomon and Sheb* (King Vidor, 1959) alongside Gina Lollobrigida at the time.

234
1963 ASTON MARTIN DB5
Sean Connery, the perfect James Bond, poses with the gadget-laden *Goldfinger* DB5; the car was a perfect match for Connery's iron-fist-in-velvet-glove film persona.

235
1964 US "STOCK" CAR
Elvis and Ann-Margaret looking utterly cool and unfazed at speed in the Nevada desert as Lucky Jackson and Rusty Martin in a sporting number in a scene from *Viva Las Vegas* (George Sidney, 1964).

236

237

236+237
1954 PORSCHE 550
James Byron Dean came to a nasty end on September 30, 1955 when he smashed his lightweight (550kg) racing Porsche near the intersection of highways 41 and 46 at Cholame, California. He was 24 years old and an up-and-coming Hollywood star. The car was just one of 78 sold to the public. One of his fellow actors in *Rebel Without a Cause* (Nicholas Ray, 1955) was Beverly Long. Later, she said of a ride in the fated car with Dean, "I felt like I was riding in a tombstone. I had never ridden in a Porsche. And when you sit down, you sit down. And you feel like you're on the ground. I had the feeling I was sitting in a coffin. It was very scary. And Jimmy drove way too fast." The Porsche was designed with a cause, to win its class at Le Mans, which it did in 1954. It also won the classic Targa Florio, driven by Umberto Maglioli, in 1956. Designed by Wilhelm Hild and bodied by Weidenhusen of Frankfurt, it had a top speed of 220km/h. It was not, despite its looks, a toy. James Dean called it his "Little Bastard." But that depended on how this fierce and brilliant little car was driven.

238
1966 AC COBRA 427
Carrol Shelby first fitted the tough, chuckable British AC Ace sports car with a mighty, high-revving, seven-litre Ford V8. The result was a blistering racing car and one of the fastest accelerating road cars of all time. Elvis got to drive one on *Spinout* (Norman Taurog, 1966). Pursued not just by other racing cars, but also by three pouting lovelies determined to race him up to the altar, Elvis stays loyal to his Cobra. The advertising blurb screamed, "It's Elvis with his foot on the gas and no brakes on the fun!!!" Hmm. The cars are great, but with songs like "Adam and Evil" and, I kid you not, "Smorgasbord", *Spinout* seems to have spun off the classic movie circuit. ACs are still made to thrill at Thames Ditton in Surrey.

GARAGE
RED CROWN
GASOLINE
ZEROLENE

240

239
1918 MODEL-T FORD
Hollywood's silent stars got a lot of laughs smashing up horseless carriages. Here's the great, deadpan comedian Buster Keaton (1895–1966) typically unfazed as the Tin Lizzie driven by Fatty Arbuckle (1887–1933) collapses in front of a Hollywood garage.

240
1954 AUSTIN J40
This is Norman Wisdom (b 1915), an English comedian popular in the 1950s who went on to become a star in Romania in later years – the brutal Communist dictator, Nicolae Ceausescu found him funny. However, he did remarkably well as a serious actor in later life. In this chase scene from *One Good Turn* (John Carstairs, 1954), Wisdom is driving an Austin Junior Forty pedal car. These delightful toys – all 32,098 of them – were made by Austin in a factory at Bargoed in South Wales by disabled coal miners, from 1949 to 1971. Originally, they cost £33 and were based on the design of the 1948 Austin A40 Devon. A working horn and headlamps were included in the price. Later, the factory made components for the long-lived BMC/Leyland/Mini A-series engine, until its closure in 1999.

241

241 + 242
1969 MINI COOPER S
The real stars of *The Italian Job* (Peter Collinson, 1969) were not Michael Caine, Noel Coward, Benny Hill or even the wonderful Fred Emney and John Le Mesurier, but the cars. There were a lot of these. They included not just a trio of brilliantly driven Cooper S Minis ricocheting through Turin, but a Lamborghini Miura, Aston Martin DB6 Volante, a pair of E-Type Jags, a Series 1 Land Rover, Fiat Dino Coupes (the Mafia), Alfa Romeo Giulias (the police) and a triple-axled Bedford Legionnaire coach that steals the final scene. Here, one of the Minis is seen interrupting a wedding, while all three are caught squealing along the *terrazo* floors of one of Turin's magnificent nineteenth century arcades. A forthcoming American remake of the film stars three of the new and much brawnier BMW Minis, although 32 have been used, and abused, in the actual making of the movie.

246

247

248
1949 MERCURY
James Dean made two cars famous – the Porsche 550 he died in and the '49 Mercury V8 he drove in the cult teenage angst classic *Rebel Without a Cause* (Nicholas Ray, 1955). The Mercury, which looked like an inverted bath tub, was actually a quick and capable car; it could top 100mph and boasted independent front suspension, so could wobble around corners better than many less sophisticated contemporaries. It was also a car that young Americans enjoyed hotting-up; the style of the car took to this well. They remain much sought-after classics today. In this publicity shot for the film, the driver's window seems the wrong profile for a Mercury, but, hell, that's the car our Jimmy drove.

249
1961 FERRARI 250GT CALIFORNIA
In *Ferris Bueller's Day Off* (John Hughes, 1986), this superb Scaglietti-bodied Ferrari appears to have a very rough time, at one point shooting through a sheet of plate glass. Not to worry. Four replicas were made for the film by Modena Design and Development, El Cajon, California. These five-litre Ford V8 replicas may not have the ultimate cachet of a multi-cog-spinning V12 Ferrari, but you would have to be exceptionally keen eyed to spot the difference on film. The Californian-built cars are not allowed to carry the prized Ferrari leaping horse badge. That's how you tell, unless you raise the hood.

250
1971 PORSCHE 917
Among the all-time great racing cars, the Porsche 917 was a formidable 250mph beast that trounced all opposition in its Le Mans heyday. Steve McQueen drove a Gulf-Porsche 917 fast and well in *Le Mans* (Lee H Katzin, 1971), his sequences, and those of other professional drivers hired for the film, cut in and out of footage of the real 1970 event when 917s came home 1-2-3. Driving in the film was full-on; one driver lost a leg when he crashed. The 917 was powered a truly Teutonic 4.5-litre (later five-litre) flat 12 that cranked out between 580bhp (1969) and 1,560bhp (1973, 917/30). The racing rule books were rewritten for 1974 to keep the 917 out of the picture and the all-conquering machine was forced to retire. Do not even attempt to watch the film unless you like fast, powerful cars. There is not much else in it.

251
1961 FORD THUNDERBIRD
This 300bhp, lipstick-red tearaway was a perfect mobile prop for Susan Sarandon and Gina Davis, stars of the feminist road movie *Thelma and Louise* (Ridley Scott, 1991). If ever a car suggests "freedom of the road", this it it.

252
1963 VOLKSWAGEN BEETLE
Car Number 53, where are you? Right on top of your head, buddy. This is Herbie in action. The Bug starred in four cute Disney films as well as a duff TV series. It was a neat idea to take a car that can barely top 80mph and turn it into a child-friendly supercar. There are some delightful parodies of other films and Herbie's antics are amusing. The four films were *The Love Bug* (1969), *Herbie Rides Again* (1974), *Herbie Goes to Monte Carlo* (1977) and *Herbie Goes Bananas* (1980).

248

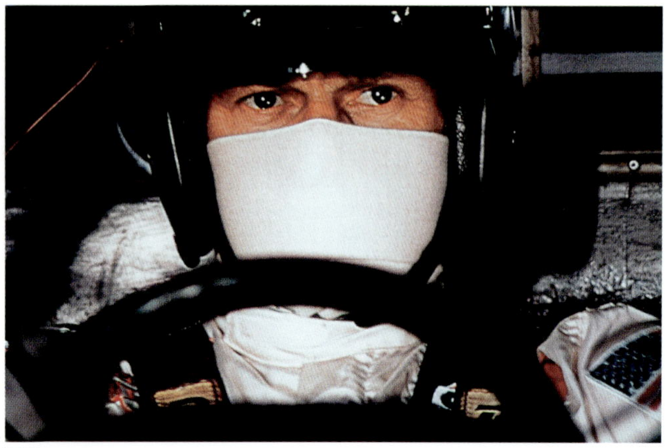
250

249

251

252

253
1966 BATMOBILE
Holy Ford, Batman! This was really the $250,000 1955 Lincoln Futura concept car in cartoon-style disguise. Originally designed by Bill Schmidt, after an encounter with a real life shark, the shark-like car was bodied by Ghia and toured the US until its retirement in 1959. It popped up again in *It Started with a Kiss* (George Marshall, 1959) starring Debbie Reynolds and Glen Ford. In 1966 it was converted – in just three weeks before filming began – into the Batmobile by George Barris for ABC's new "Batman" series. Batman (Adam West) and Robin (Burt Ward) camped about happily in the intentionally funny Bam! Pow! Pop! shows, and made the Batmobile a star with children everywhere. A number of plastic replicas were made during the series, which ran until 1968; a custom car expert, George Barris owns the original, steel-bodied car.

254

254
1962 VOLVO P1800
Leslie Charteris's fictional hero, Simon Templar aka The Saint, drove an equally fictional Hirondel. When the story was made into a TV series, "The Saint", in 1962, Jaguar refused an E-Type, much to its later regret – the programme was extremely popular and ran for years – and Volvo drove in with a white 115bhp P1800. This was never the fastest or most dynamic sports car; in fact it was hardly a sports car at all. But it looked good. Four cars were supplied over the years, including one for Roger Moore's personal use. The 1964 car, 77 GYL, is on permanent show at the Cars of the Stars Museum, Keswick, England. For the first two years of its life, the P1800 was assembled by Jensen in West Bromwich, England; production was transferred to Gothenburg in 1963.

255
1975 FORD GRAN TORINO
The cops-chase-robbers "Starsky and Hutch" TV series starring David Soul and Paul Glaser had such a cult following that, for 1976, Ford issued a special line of 1,000 tomato red and white-striped Starsky and Hutch Gran Torinos. A case of life imitating art. The TV car had a 400 cubic inch V8 under the hood and was fast.

256
1969 DODGE CHARGER
Bo, Luke and Daisy's mount in the "Dukes of Hazzard" was General Lee, a customized, bright orange '69 Dodge Charger that did a lot of charging around Hazzard County, set, fictionally, in Georgia. Filming of the car chase TV series was split between Georgia and California. Great fun. Fun car.

255

256

257

258

257
1965 ROLLS-ROYCE PHANTOM V
Delivered new to John Lennon in June 1965, this Mulliner Park Ward bodied Phantom V was originally painted glossy black. In April 1967, while making "Sergeant Pepper's Lonely Heart's Club Band", Lennon took the car to J P Fallon Ltd, a Chertsey coachbuilder. For about £2,000, the car was painted in swirls and swags of flowers by a team of Dutch gypsy artists called The Fool. Lennon took the car to New York in 1970 where he lent it, variously, to the Rolling Stones, Moody Blues and Bob Dylan. Donated to the Cooper-Hewitt Museum, New York, in 1977 in exchange for a $225,000 tax credit, the car is now preserved in the Ottowa Museum of Science and Technology.

258
1960s ROLLS-ROYCES
A trio of Rolls-Royces outside Elstree Studios in February 1969, where Tom Jones was recording his ABC-TV series "This is Tom Jones". Jones (left) sits astride his 1966 "Chinese Eyes" Silver Cloud III (you could get away with labels like that in the politically incorrect 1960s) and the 1966 Silver Shadow of his manager, Gordon Mills. On the right with his 1965 Phantom V is singer Engelbert Humperdinck. The boys had done good.

259
1966 PONTIAC GTO
Here they come, driving down the street, getting funniest looks from everyone that they meet. No surprise when four gurning young men are leaning out of a highly customized Pontiac. This car, the Monkeemobile, was one of two created by Dean Jeffries. It featured three rows of seats, a parachute and, originally, a blower that allowed the cars to perform spectacular wheelies. Wisely, this was removed. The Monkees – Mike Nesmith, Peter Tork, Micky Dolenz and Davy Jones – were each given a stock '66 GTO by Pontiac. Nice work if you can get it. One of the Monkeemobiles is now owned by George Barris, creator of the Batmobile.

259